"Head First did it again. The ability to make the reader understand, despite tricky topics, really shines through in *Head First 2D Geometry*! The way the information is presented and organized makes learning cohesive and easy. Coming from someone who has struggled with many aspects of math in the past, this book helps you understand the basics and build on them. I wish I had this book when I was taking Geometry!"

> **— Amanda Borcky**
> **College student**

"*Head First 2D Geometry* is a clearly written guide to learn about two-dimensional shapes. The thorough explanations of the material are adequate for both a first-time student and one needing a quick review. The 'hands on' approach gives a richer understanding of the material than would otherwise be obtained from a traditional textbook."

> **— Ariana Anderson**
> **Statistician at UCLA's Center for Cognitive Neuroscience**

"*Head First 2D Geometry* helps you learn that plane geometry doesn't have to be *plain* geometry. This book lets you see that geometry is not only in the classroom, it is all around you and a part of your everyday life."

> **— Herbert Tracey**
> **Instructor of mathematical sciences at Loyola University Maryland and**
> **former department chair of mathematics at Hereford High School**

"*Head First 2D Geometry* is clear and readable, while other textbooks drag students through a thicket of academic jargon. Head First has interesting examples, fun design, and a conversational style that the textbook industry would do well to emulate."

> **— Dan Meyer**
> **High school math teacher and recipient of Cable in the Classroom's**
> **Leader in Learning award**

"*Head First 2D Geometry* grabs your attention with inventive and clever applied problems. It pursues thorough solutions with persistence and energy. There is one character who appears throughout the book and delights me—a serious, seemingly humorless girl who suspects the authors are trying to get away with inconsistency and poor logic. They always praise her questions and give in to her demands that they level with her."

> **— David Meyer**
> **Retired college and high school math teacher**

Praise for other Head First books

"*Head First Algebra* is a clear, easy-to-understand method to learn a subject that many people find intimidating. Because of its somewhat irreverent attitude in presenting mathematical topics for beginners, this book inspires students to learn Algebra at a depth they might have otherwise thought unachievable."

> **— Ariana Anderson**
> **Statistician at UCLA's Center for Cognitive Neuroscience**

"The way *Head First Algebra* presents information is so conversational and intriguing it helps in the learning process. It truly feels like you're having a conversation with the author."

> **— Amanda Borcky**
> **College student**

"*Head First Algebra* has got to be the best book out there for learning basic algebra. It's genuinely entertaining."

> **— Dawn Griffiths**
> **Author, *Head First Statistics***

"*Head First Algebra* is an engaging read. The book does a fantastic job of explaining concepts and taking the reader step-by-step through solving problems. The problems were challenging and applicable to everyday life."

> **— Shannon Stewart**
> **Math teacher**

"*Head First Algebra* is driven by excellent examples from the world in which students live. No trains leaving from the same station at the same time moving in opposite directions. The authors anticipate well the questions that arise in students' minds and answer them in a timely manner. A very readable look at the topics encountered in Algebra 1."

> **— Herbert Tracey**
> **Instructor of mathematical sciences, Loyola University Maryland**

"If you want to learn some physics, but you think it's too difficult, buy *Head First Physics*! It will probably help, and if it doesn't, you can always use it as a doorstop or hamster bedding or something. I wish I had a copy of this book when I was teaching physics."

> **— John Allister**
> **Physics teacher**

Praise for other Head First books

"*Head First Physics* has achieved the impossible—a serious textbook that makes physics fun. Students all over will be thinking like a physicist!"

> — **Georgia Gale Grant**
> **Freelance science writer, communicator, and broadcaster**

"Great graphics, clear explanations, and some crazy real-world problems to solve! *Head First Physics* is full of strategies and tips to attack problems. It encourages a team approach that's so essential in today's work world."

> — **Diane Jaquith**
> **High school physics, chemistry, and physical science teacher**

"*Head First Physics* is an outstandingly good teacher masquerading as a physics book! You never feel fazed if you don't quite understand something the first time because you know it will be explained again in a different way and then repeated and reinforced."

> — **Marion Long**
> **Teacher**

"Dawn Griffiths has split some very complicated concepts into much smaller, less frightening bits of stuff that real-life people will find very easy to digest. *Head First Statistics* has lots of graphics and photos that make the material very approachable, and I have developed quite a crush on the attractive lady model who is asking about gumballs on page 458."

> — **Bruce Frey**
> **Author, *Statistics Hacks***

"*Head First Statistics* is an intuitive way to understand statistics using simple, real-life examples that make learning fun and natural."

> — **Michael Prerau**
> **Computational neuroscientist and statistics instructor, Boston University**

"Thought Head First was just for computer nerds? Try the brain-friendly way with *Head First Statistics* and you'll change your mind. It really works."

> — **Andy Parker**

"Down with dull statistics books! Even my cat liked *Head First Statistics*."

> — **Cary Collett**

Other related books from O'Reilly

Statistics in a Nutshell

Statistics Hacks

Mind Hacks

Mind Performance Hacks

Your Brain: The Missing Manual

Other books in O'Reilly's *Head First* series

Head First C#

Head First Java

Head First Object-Oriented Analysis and Design (OOA&D)

Head First HTML with CSS and XHTML

Head First Design Patterns

Head First Servlets and JSP

Head First EJB

Head First SQL

Head First Software Development

Head First JavaScript

Head First Physics

Head First Statistics

Head First Ajax

Head First Rails

Head First Algebra

Head First PHP & MySQL

Head First PMP

Head First Web Design

Head First Networking

Head First Programming

Head First 2D Geometry

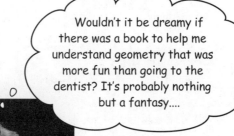

Wouldn't it be dreamy if there was a book to help me understand geometry that was more fun than going to the dentist? It's probably nothing but a fantasy....

Lindsey Fallow
Dawn Griffiths

O'REILLY®

Beijing • Cambridge • Köln • Sebastopol • Taipei • Tokyo

Head First 2D Geometry

by Lindsey Fallow and Dawn Griffiths

516.9
Fal

Published by O'Reilly Media, Inc., 1005 Gravenstein Highway North, Sebastopol, CA 95472.

O'Reilly Media books may be purchased for educational, business, or sales promotional use. Online editions are also available for most titles (*http://my.safaribooksonline.com*). For more information, contact our corporate/institutional sales department: (800) 998-9938 or *corporate@oreilly.com*.

Series Creators:	Kathy Sierra, Bert Bates
Series Editor:	Brett D. McLaughlin
Editor:	Courtney Nash
Design Editor:	Dawn Griffiths
Cover Designer:	Karen Montgomery
Production Editor:	Rachel Monaghan
Indexer:	Angela Howard
Proofreader:	Nancy Reinhardt
Page Viewers:	Badger, Helen, Joe, David, and Carl

David

Printing History:

Helen

November 2009: First Edition.

Badger

Joe

Carl

RepKover.
This book uses Repkover,™ a durable and flexible lay-flat binding.

ISBN: 978-0-596-80833-4

[M]

To Mum and Dad for buying me construction kits. To my fantastic Yorkshire family for endless support, humour, and psychotherapy— I love you even more than I love triangles. And to triangles and sheep, for making the world a fascinating place to be.

—Lindsey

To David, Mum, Dad, and Carl for their ongoing love and support. Also in loving memory of Peter Lancaster Walker, an unsung hero who made so many things possible.

—Dawn

Lindsey

↟ Ruby the boxer.

Dawn

Lindsey Fallow is a self-confessed geek who has spent the past decade exploring science and technology as a writer, software developer, and TV presenter.

After earning her undergraduate degree in manufacturing engineering, she fronted a science show for 8–12-year-olds on Disney, and went on to become a reporter and associate producer for *Tomorrow's World* (the BBC's #1 prime-time UK science and technology show) from 1998–2002.

She's stood on the top of the Golden Gate bridge, fed sharks, filmed brain surgery, flown in military planes, and been bitten by a baby tiger, but is the most excited by far when her 14-year-old stepson "gets" his math homework.

She is an avid fan of the Head First series and can't quite believe she's actually written one.

Lindz claims that if she were a superhero, her superpower would be tesselating. When she's not working, she likes to spend time with her super-lovely partner Helen, dabble in sheep farming, play with her boxer dog, Ruby, rock the drums on *Guitar Hero*, and walk in the wilderness.

Dawn Griffiths started life as a mathematician at a top UK university where she was awarded a first-class honours degree in mathematics. She went on to pursue a career in software development, and she currently combines IT consultancy with writing, editing, and mathematics.

Dawn is the author of *Head First Statistics*, and has also worked on a host of other books in the series, from Networking to Programming.

When Dawn's not working on Head First books, you'll find her honing her Tai Chi skills, making bobbin lace, or cooking. She hasn't yet mastered the art of doing all three at the same time. She also enjoys traveling, and spending time with her wonderful husband, David.

Dawn has a theory that *Head First Bobbin Lacemaking* might prove to be a big cult hit, but she suspects that Brett might disagree.

Table of Contents (Summary)

	Intro	xvii
1	Finding missing angles: *Reading between the lines*	1
2	Similarity and congruence: *Shrink to fit*	49
3	The Pythagorean Theorem: *All the right angles*	103
4	Triangle properties: *Between a rock show and a triangular place*	149
5	Circles: *Going round and round*	205
6	Quadrilaterals: *It's hip to be square*	235
7	Regular polygons: *It's all shaping up*	273

Table of Contents (the real thing)

Intro

Your brain on Geometry. Here *you* are trying to *learn* something, while here your *brain* is doing you a favor by making sure the learning doesn't *stick*. Your brain's thinking, "Better leave room for more important things, like which wild animals to avoid and whether naked snowboarding is a bad idea." So, how *do* you trick your brain into thinking that your life depends on knowing about triangles and circles and the Pythagorean Theorem?

Who is this book for?	xviii
We know what you're thinking	xix
Metacognition: thinking about thinking	xxi
Here's what YOU can do to bend your brain into submission	xxiii
Read me	xxiv
The technical review team	xxvi
Acknowledgments	xxvii

finding missing angles

Reading between the lines

Ever get the feeling there's something they're not telling you?

If you want to master the **real world**, you need to get geometry. It's a set of **tools** for turning a little bit of information into a ***complete picture***. Whether you want to design something, build something or find out how a situation really went down, geometry can make sure you've always got **the lowdown**. So if you want to keep in the loop, grab your hat, pack your pencil, and join us on the bus to Geometryville.

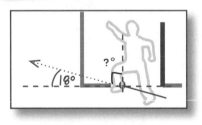

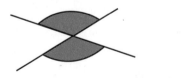

There's been a homicide	2
In the ballistics lab you've got to cover all the angles	3
Do the angles between Benny, Micky, and the bullet match up?	4
Right angles aren't always marked with numbers	6
Angles can be made up of other, smaller angles	7
Complementary angles always add up to a right angle (90°)	9
Right angles often come in pairs	11
Angles on a straight line add up to 180°	14
Pairs of angles that add up to 180° are called supplementary angles	17
Vertical angles are always equal	19
The corner angles of a triangle always add up to a straight line	20
Find one more angle to crack the case	21
Something doesn't add up!	23
If it doesn't all add up, then something isn't as it seems	24
You've proved that Benny couldn't have shot Micky!	25
We've got a new sketch—now for a new ballistics report	26
We need a new theory	27
Work out what you need to know	29
Tick marks indicate equal angles	30
Use what you know to find what you don't know	31
The angles of a four-sided shape add up to 360°	35
Parallel lines are lines at exactly the same angle	39
Parallel lines often come with helpful angle shortcuts	40
Great work—you cracked the case!	44
Your Geometry Toolbox	46

similarity and congruence

Shrink to fit

2

Sometimes, size does matter.

Ever drawn or built something and then found out it's the **wrong size**? Or made something just perfect and wanted to *recreate it exactly?* You need *Similarity* and *Congruence*: the **time-saving techniques for duplicating** your designs smaller, bigger, or exactly the same size. Nobody likes doing the same work over—and with similarity and congruence, you'll *never have to repeat an angle calculation again.*

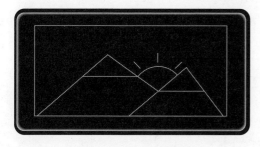

Welcome to myPod! You're hired	50
Liz wants you to etch her phone	51
The designer noted some of the details	53
The design tells us that some triangles are repeated	54
Similar triangles don't just look the same	58
To use similarity, you need to be able to spot it	61
You can spot similar triangles based on just two angles	62
Employee of the month already?	66
You sketch it—we'll etch it!	67
Fire up the etcher!	68
The boss isn't happy, but at least you're not fired…	69
It's a problem of scale…	70
Complex shapes can be similar, too	73
You sketch it—we'll etch it (to fit)	77
Liz is back with a special request	78
Similar shapes that are the same size are congruent	81
Use what you know to find what you don't know	83
Ratios can be more useful than sizes	93
Ratios need to be consistent	96
Your new design ROCKS!	99
Your Geometry Toolbox	100

table of contents

the pythagorean theorem

All the right angles

Sometimes, you really need to get things straight.

Ever tried to eat at a wobbly table? Annoying, isn't it? There is an alternative to shoving screwed-up paper under the table leg though: use the Pythagorean Theorem to make sure your designs are **dead straight** and not just *quite straight*. Once you know this pattern, you'll be able to **spot and create right angles** that are **perfect every time.** Nobody likes to pick spaghetti out of their lap, and with the **Pythagorean Theorem**, you don't have to.

Giant construction-kit skate ramps	104
Standard-sized-quick-assembly-what?!?	105
The ramps must have perpendicular uprights	106
You can use accurate construction to test ramp designs on paper	108
Not all lengths make a right triangle	115
You can explore a geometry problem in different ways	116
In geometry, the rules are the rules	118
Any good jump has some similar scaled cousins	121
The lengths of the sides are linked by a pattern	126
The square of the longest side is equal to the squares of the other two sides added together	130
The Pythagorean Theorem: $a^2 + b^2 = c^2$	131
Using Kwik-klik skate ramps is definitely the right angle!	137
A longer rope swings further and lower	140
So, how far can you swing on a six-meter rope?	142
Your rope swing is perfect	147
Your Geometry Toolbox	148

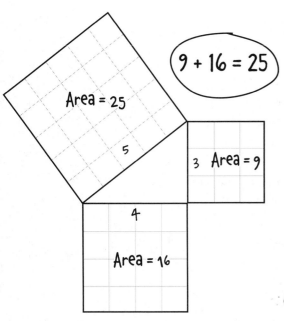

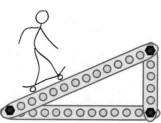

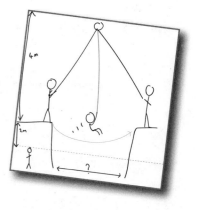

triangle properties

Between a rock show and a triangular place

Ever had that sinking feeling that you've made a bad decision?

In the real world, *choices can be complex*, and wrong decisions can cost you **money** and **time**. Many solutions aren't always straightforward: even in geometry, bigger doesn't always mean better—it might not even mean longer. *So what should you do?* The good news is that you can *combine your triangle tools* to **make great decisions** even when it seems like you don't have the right information to answer the question.

Everyone loves organizing a rock festival	150
First we need to pick a venue	151
Fencing costs money	153
Does a bigger perimeter mean a bigger area?	155
How many people can each venue hold?	156
A triangle fits inside a bounding rectangle	157
The area of a triangle = 1/2 base × height	163
You've got $11,250 to spend	168
All speakers are not created equal	170
So what are you looking for in your speakers?	171
The ideal speakers are wider and longer than the venue… but only by a little	173
100m will do, but can you rent the 60° speaker?	175
The 60° speakers are spot on	178
All that's left is to pick a spot for the drinks stall	181
A triangle has more than one center	182
The center of a triangle can be outside the triangle	186
Let's put the drink stall at the centroid	187
The rock festival is ready!	189
The people behind the drinks stall won't see the stage…	190
You need a screen for less than $1,440	192
Will the special offer screen still do the job?	193
You can find area from sides using Hero's formula	196
Hero's formula and "1/2 base × height" work together	198
The rock festival is gonna…rock!	201
Your Geometry Toolbox	202

circles

Going round and round

OK, life doesn't have to be so straight after all!

There's no need to reinvent the wheel, but aren't you glad you're able to use it? From cars to rollercoasters, many of the **most important solutions** to life's problems rely on *circles* to get the job done. Free yourself from straight edges and pointy corners—there's no end to the curvy possibilities once you master **circumference**, **arcs**, and **sectors**.

It's not just pizza—it's war!	206
How does MegaSlice's deal measure up?	207
The diameter of a circle is twice its radius	208
How do slices compare to whole pizzas?	209
Sectors of a circle have angles totaling 360°	210
MegaSlice's $10 deal is a con!	211
Pepperoni crust pizza—but at what price?	212
The pepperoni perimeter is 3 (and a bit) times diameter	214
Mario wants to put your pepperoni crust pricing formula to the test	217
The customers are always fussy	219
An arc is a section of the circumference	221
Mario's business is booming!	222
But MegaSlice is at it again...	223
We need to find the area of the two pizza deals	224
Each sector (slice) is a triangle (kind of)	226
Area of a circle = πr^2	231
Mario's pizza is here to stay	233
Your Geometry Toolbox	234

Π ← Pi

50 bits of pepperoni

$\dfrac{50}{16} = 3.125$

quadrilaterals

It's hip to be square

Maybe three isn't the (only) magic number.

The world isn't just made up of triangles and circles. Wherever you look, you'll see **quadrilaterals**, shapes with four straight sides. Knowing your way 'round the quad family can save you a lot of time and effort. Whether it's **area**, **perimeter**, or **angles** you're after, there are *shortcuts galore* that you can **use to your advantage**. Keep reading, and we'll give you the lowdown.

6

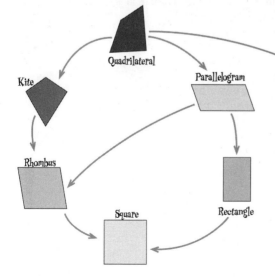

~~~~~~~~~~Edward's Lawn Service~~~~~~~~~~

Lawn cutting cost - $0.05 per square meter

Lawn edging cost - $0.10 per meter

(Payable weekly)

Edward's Lawn Service needs your help	236
Your first lawn	237
The lawn is a parallelogram	238
Let's split the parallelogram	239
Business is booming!	241
If you don't like what you're given, change it	245
But people are upset with Ed's prices…	247
Let's compare the two lawns	248
The lawns need edging, too	249
Same shape, different perimeters	250
Edward changed his rates…	252
…and the customers keep flooding in	253
Use diagonals to find the area of the kite	257
Landowners, unite	260
There are some familiar things about this shape	262
Calculate trapezoid area using base length and height	264
The quadrilateral family tree	268
You've entered the big league	271
Your Geometry Toolbox	272

Quadrilateral

Kite

Parallelogram

Trapezoid

Rhombus

Isosceles Trapezoid

Square

Rectangle

regular polygons
It's all shaping up

Want to have it your way? Life's full of compromises, but you don't have to be restricted to triangles, squares, and circles. **Regular polygons** give you the **flexibility** to demand exactly the shape you need. But don't think that means learning a list of new formulas: *you can treat 6-, 16-, and 60-sided shapes the same*. So, whether it's for your own creative project, some required homework that's due tomorrow, or the demands of an important client, you'll have the tools to **deliver exactly what you want**.

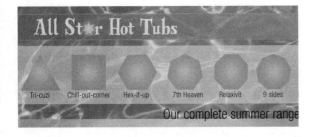

All Star Hot Tubs

Tri-cuzi   Chill-out-corner   Hex-it-up   7th Heaven   Relaxiv8   9 sides

Our complete summer range

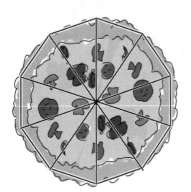

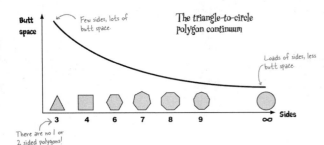

Butt space

Few sides, lots of butt space.

The triangle-to-circle polygon continuum

Loads of sides, less butt space.

Sides

3   4   6   7   8   9   ∞

There are no 1 or 2 sided polygons!

We need to choose a hot tub	274
All the hot tubs are regular polygons	275
Regular polygons have equal sides and angles	276
Butt-space is all about perimeter	277
Is 3 cubic meters of water a lot or a little?	278
Hot tub volume is area × depth	280
The hot tub's area must be 6m²	282
Which hot tub shape gives the most butt-space?	283
Work backward from area to find butt-space	284
Is 19.6 butts a lot or a little?	287
The square tub beats the circle tub	288
Two tubs down, five to go	289
You've found the formula for the area of an equilateral triangle	293
Keep track of complex comparisons with a table	296
Chop the polygons into triangles	302
What do we need to know about the polygon triangles?	303
The circles give us the properties we need	306
Polygon area = 1/2 perimeter × apothem	309
More sides = fewer butts	310
Rock stars—high maintenance?	311
Great tub choice!	313
But what about dimensions?	315
It's time to relax in the hot tub!	317
Your Geometry Toolbox	318
Leaving town…	319
It's been great having you here in Geometryville!	319

how to use this book

Intro

In this section, we answer the burning question:
"So why <u>DID</u> they put that in a geometry book?"

Who is this book for?

If you can answer "yes" to all of these…

 Are you already pretty comfortable with algebra? ← *If not, check out Head First Algebra first!*

 Do you want to **learn**, **understand**, **remember**, and *apply* geometry concepts, and not just memorize formulas?

③ Do you prefer **fun, casual conversation** to dry, **dull, school lectures**?

…this book is for you.

Who should probably back away from this book?

If you can answer "yes" to any of these…

① Are you **still struggling** with solving for unknowns in algebra? ← *If you can solve $3(x + 4) = 21$, you'll be fine*

② Are you afraid of sketching, drawing, and using your hands to figure things out?

③ Are you someone who'd rather just plug stuff into calculators or have someone give you the answers? Do you believe that a math book can't be serious if there's a rock concert in it?

…this book is not for you.

[Note from marketing: this book is for anyone with a credit card. Or cash. Cash is nice, too. —Ed]

We know what you're thinking.

"How can *this* be a serious geometry book?"

"What's with all the graphics?"

"Can I actually *learn* it this way?"

And we know what your *brain* is thinking.

Your brain craves novelty. It's always searching, scanning, *waiting* for something unusual. It was built that way, and it helps you stay alive.

So what does your brain do with all the routine, ordinary, normal things you encounter? Everything it *can* to stop them from interfering with the brain's *real* job—recording things that *matter*. It doesn't bother saving the boring things; they never make it past the "this is obviously not important" filter.

How does your brain *know* what's important? Suppose you're out for a day hike and a tiger jumps in front of you, what happens inside your head and body?

Neurons fire. Emotions crank up. *Chemicals surge.*

And that's how your brain knows....

This must be important! Don't forget it!

But imagine you're at home, or in a library. It's a safe, warm, tiger-free zone. You're studying. Getting ready for an exam. Or trying to learn some tough math thing that your teacher is going to test you on tomorrow.

Just one problem. Your brain's trying to do you a big favor. It's trying to make sure that this *obviously* non-important content doesn't clutter up scarce resources. Resources that are better spent storing the really *big* things. Like tigers. Like the danger of fire. Like how you should never again snowboard in shorts.

And there's no simple way to tell your brain, "Hey brain, thank you very much, but no matter how dull this book is, and how little I'm registering on the emotional Richter scale right now, I really *do* want you to keep this stuff around."

Your brain thinks THIS is important.

Great. Only 330 more dull, dry, boring pages.

Your brain thinks THIS isn't worth saving.

We think of a "Head First" reader as a <u>learner</u>.

So what does it take to *learn* something? First, you have to *get* it, then make sure you don't *forget* it. It's not about pushing facts into your head. Based on the latest research in cognitive science, neurobiology, and educational psychology, *learning* takes a lot more than text on a page. We know what turns your brain on.

Some of the Head First learning principles:

Make it visual. Images are far more memorable than words alone, and make learning much more effective (up to 89% improvement in recall and transfer studies). It also makes things more understandable.

Put the words within or near the graphics they relate to, rather than on the bottom or on another page, and learners will be up to *twice* as likely to solve problems related to the content.

> OMG! Come check out this email—this is so cool!

Use a conversational and personalized style. In recent studies, students performed up to 40% better on post-learning tests if the content spoke directly to the reader, using a first-person, conversational style rather than taking a formal tone. Tell stories instead of lecturing. Use casual language. Don't take yourself too seriously. Which would *you* pay more attention to: a stimulating dinner party companion, or a lecture?

Get the learner to think more deeply. In other words, unless you actively flex your neurons, nothing much happens in your head. A reader has to be motivated, engaged, curious, and inspired to solve problems, draw conclusions, and generate new knowledge. And for that, you need challenges, exercises, and thought-provoking questions, and activities that involve both sides of the brain and multiple senses.

Get—and keep—the reader's attention. We've all had the "I really want to learn this but I can't stay awake past page one" experience. Your brain pays attention to things that are out of the ordinary, interesting, strange, eye-catching, unexpected. Learning a new, tough, technical topic doesn't have to be boring. Your brain will learn much more quickly if it's not.

> Dude, we sold ALL the tickets! Are you ready to ROCK?

Touch their emotions. We now know that your ability to remember something is largely dependent on its emotional content. You remember what you care about. You remember when you *feel* something. No, we're not talking heart-wrenching stories about a boy and his dog. We're talking emotions like surprise, curiosity, fun, "what the...?" , and the feeling of "I Rule!" that comes when you solve a puzzle, learn something everybody else thinks is hard, or realize you know something that "I'm more technical than thou" Bob from engineering *doesn't*.

Metacognition: thinking about thinking

If you really want to learn, and you want to learn more quickly and more deeply, pay attention to how you pay attention. Think about how you think. Learn how you learn.

Most of us did not take courses on metacognition or learning theory when we were growing up. We were *expected* to learn, but rarely *taught* to learn.

I wonder how I can trick my brain into remembering this stuff....

But we assume that if you're holding this book, you really want (or need) to learn about geometry. And you probably don't want to spend a lot of time. And since you're going to have to use this stuff in the future, you need to *remember* what you read. And for that, you've got to *understand* it. To get the most from this book, or *any* book or learning experience, take responsibility for your brain. Your brain on *geometry*.

The trick is to get your brain to see the new material you're learning as Really Important. Crucial to your well-being. As important as a tiger. Otherwise, you're in for a constant battle, with your brain doing its best to keep the new content from sticking.

So just how *DO* you get your brain to think that geometry is a hungry tiger?

There's the slow, tedious way, or the faster, more effective way. The slow way is about sheer repetition. You obviously know that you *are* able to learn and remember even the dullest of topics if you keep pounding the same thing into your brain. With enough repetition, your brain says, "This doesn't *feel* important to him, but he keeps looking at the same thing *over* and *over* and *over*, so I suppose it must be."

The faster way is to do **anything that increases brain activity,** especially different *types* of brain activity. The things on the previous page are a big part of the solution, and they're all things that have been proven to help your brain work in your favor. For example, studies show that putting words *within* the pictures they describe (as opposed to somewhere else in the page, like a caption or in the body text) causes your brain to try to makes sense of how the words and picture relate, and this causes more neurons to fire. More neurons firing = more chances for your brain to *get* that this is something worth paying attention to, and possibly recording.

A conversational style helps because people tend to pay more attention when they perceive that they're in a conversation, since they're expected to follow along and hold up their end. The amazing thing is, your brain doesn't necessarily *care* that the "conversation" is between you and a book! On the other hand, if the writing style is formal and dry, your brain perceives it the same way you experience being lectured to while sitting in a roomful of passive attendees. No need to stay awake.

But pictures and conversational style are just the beginning.

Here's what WE did:

We used **pictures**, because your brain is tuned for visuals, not text. As far as your brain's concerned, a picture really *is* worth a thousand words. And when text and pictures work together, we embedded the text *in* the pictures because your brain works more effectively when the text is *within* the thing the text refers to, as opposed to in a caption or buried in the text somewhere.

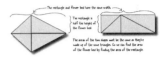

We used **redundancy**, saying the same thing in *different* ways and with different media types, and *multiple senses*, to increase the chance that the content gets coded into more than one area of your brain.

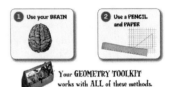

We used concepts and pictures in **unexpected** ways because your brain is tuned for novelty, and we used pictures and ideas with at least *some* **emotional** *content*, because your brain is tuned to pay attention to the biochemistry of emotions. That which causes you to *feel* something is more likely to be remembered, even if that feeling is nothing more than a little **humor**, **surprise**, or **interest.**

We used a personalized, **conversational style**, because your brain is tuned to pay more attention when it believes you're in a conversation than if it thinks you're passively listening to a presentation. Your brain does this even when you're *reading*.

We included loads of **activities**, because your brain is tuned to learn and remember more when you **do** things than when you *read* about things. And we made the exercises challenging-yet-do-able, because that's what most people prefer.

We used **multiple learning styles**, because *you* might prefer step-by-step procedures, while someone else wants to understand the big picture first, and someone else just wants to see an example. But regardless of your own learning preference, *everyone* benefits from seeing the same content represented in multiple ways.

We include content for **both sides of your brain**, because the more of your brain you engage, the more likely you are to learn and remember, and the longer you can stay focused. Since working one side of the brain often means giving the other side a chance to rest, you can be more productive at learning for a longer period of time.

And we included **stories** and exercises that present **more than one point of view,** because your brain is tuned to learn more deeply when it's forced to make evaluations and judgments.

We included **challenges**, with exercises, and by asking **questions** that don't always have a straight answer, because your brain is tuned to learn and remember when it has to *work* at something. Think about it—you can't get your *body* in shape just by *watching* people at the gym. But we did our best to make sure that when you're working hard, it's on the *right* things. That **you're not spending one extra dendrite** processing a hard-to-understand example, or parsing difficult, jargon-laden, or overly terse text.

We used **people**. In stories, examples, pictures, etc., because, well, because *you're* a person. And your brain pays more attention to *people* than it does to *things*.

Here's what YOU can do to bend your brain into submission

So, we did our part. The rest is up to you. These tips are a starting point; listen to your brain and figure out what works for you and what doesn't. Try new things.

Cut this out and stick it on your refrigerator.

-----

① Slow down. The more you understand, the less you have to memorize.

Don't just *read*. Stop and think. When the book asks you a question, don't just skip to the answer. Imagine that someone really *is* asking the question. The more deeply you force your brain to think, the better chance you have of learning and remembering.

② Do the exercises. Write your own notes.

We put them in, but if we did them for you, that would be like having someone else do your workouts for you. And don't just *look* at the exercises. **Use a pencil.** There's plenty of evidence that physical activity *while* learning can increase the learning.

③ Read the "There are No Dumb Questions"

That means all of them. They're not optional sidebars—*they're part of the core content!* Don't skip them.

④ Make this the last thing you read before bed. Or at least the last challenging thing.

Part of the learning (especially the transfer to long-term memory) happens *after* you put the book down. Your brain needs time on its own, to do more processing. If you put in something new during that processing time, some of what you just learned will be lost.

⑤ Drink water. Lots of it.

Your brain works best in a nice bath of fluid. Dehydration (which can happen before you ever feel thirsty) decreases cognitive function.

⑥ Talk about it. Out loud.

Speaking activates a different part of the brain. If you're trying to understand something, or increase your chance of remembering it later, say it out loud. Better still, try to explain it out loud to someone else. You'll learn more quickly, and you might uncover ideas you hadn't known were there when you were reading about it.

⑦ Listen to your brain.

Pay attention to whether your brain is getting overloaded. If you find yourself starting to skim the surface or forget what you just read, it's time for a break. Once you go past a certain point, you won't learn faster by trying to shove more in, and you might even hurt the process.

⑧ Feel something!

Your brain needs to know that this *matters*. Get involved with the stories. Make up your own captions for the photos. Groaning over a bad joke is *still* better than feeling nothing at all.

⑨ Create something!

Pick up a model kit or some wood and tools and make something really cool! Or work out something you will build one day when you have the time and money. All you need is a pencil and a problem to solve…a problem that might benefit from using the tools and techniques you're studying to get geometry.

Read Me

This is a learning experience, not a reference book. We deliberately stripped out everything that might get in the way of learning whatever it is we're working on at that point in the book. And the first time through, you need to begin at the beginning, because the book makes assumptions about what you've already seen and learned.

We don't follow a regular school syllabus.

We couldn't cover every single element of the syllabus so we paid attention to what questions our own brains were asking, asked students what they found tricky, and we included extra things which allow you to find patterns that link the learning together because *your brain loves patterns.*

So, if you're going to need to pass a test, then you'll also need a reference book that covers the syllabus for that test, but don't worry. We've picked out the trickiest and most interesting parts in this book, and we've emphasized *understanding* geometry so you should be in great shape to slot those extra details into place quickly.

We don't drag you through formal proofs of new concepts.

If you're doing high school geometry you'll probably be familiar with—and possibly terrified of—geometry proofs. *There are no formal proofs in this book.* We believe that, for most people, proofs make learning geometry harder than it needs to be. Instead, we've used visual exercises to explore patterns and general rules in ways that we are confident that you'll remember and even be able to show other people.

We're working on another book in this mini-series that will handle all that formal logic and proof stuff, but for now you're in great shape if you understand geometry in the real world first.

This is just about two-dimensional (2D) geometry.

We promise it's not just so we can sell you another book called *Head First 3D Geometry* soon! We've covered many of the most important techniques you'll use when working in two dimensions, so you're all set for exploring further dimensions at the end of this book. In fact, we've even snuck in a couple of 3D problems that you can work in 2D, because geometry is about *solving interesting problems in the real world,* not just on paper.

We use plain English and not geometry jargon.

We believe your brain needs to see what something is, and figure out *why you would even care about it*, before you can give it an unfamiliar label. We do use the geometry jargon you'll need to know for tests from time to time, but not until we're sure you'll know what we're talking about. We encourage you to *use real words to describe patterns* and not sweat the official formulas too much.

We don't consider this to be the end of our conversation with you.

Come and talk to us at *www.headfirstlabs.com/geometry*. If there's something we didn't cover that's really puzzling you, then throw us a question and we'll see if we can help you with a Head First style way of figuring it out. Of course we won't do the work for you, which is why...

...the activities are NOT optional.

The exercises and activities are not add-ons; they're part of the core content of the book. Some of them are to help with memory, some are for understanding, and some will help you apply what you've learned. ***Don't skip the exercises.*** The geometry investigations are particularly important; they'll help you discover how your brain likes to figure stuff out—known as logic in formal geometry proofs that come much later (not in this book). They also give your brain a chance to hook in to the geometry that is all around you in the real world.

The redundancy is intentional and important.

One distinct difference in a Head First book is that we want you to *really* get it. And we want you to finish the book remembering what you've learned. Most reference books don't have retention and recall as a goal, but this book is about *learning*, so you'll see some of the same concepts come up more than once.

The Brain Power exercises don't have answers.

For some of them, there is no right answer, and for others, part of the learning experience of the Brain Power activities is for you to decide if and when your answers are right. In some of the Brain Power exercises, you will find hints to point you in the right direction.

The technical review team

Amanda Borcky

Herbert Tracey

Ariana Anderson

David Myers

Technical Reviewers:

For this book we had an amazing group, many of whom have reviewed other Head First books in the past. They did a fantastic job, and we're really grateful that they keep coming back for more!

Amanda Borcky is a student at Virginia Tech in Blacksburg, Virginia. She is studying nutrition with plans of getting a second degree in nursing. This is her second time reviewing for the Head First series.

David Myers taught college and high school math for 36 years. Mostly for fun, he collaborated on several math and programming textbooks in the '80s and '90s. Since retiring in 2006 from a long tenure at The Winsor School in Boston, MA, he has been delighted to start a new completely-for-fun career as a volunteer at his Quaker Meeting and in prison-related activities.

Ariana Anderson is a statistician working on "reading" brain scans at the Center for Cognitive Neuroscience at UCLA. She got her PhD from UCLA and her bachelor's from UCLA, but was forced to go to high school elsewhere.

Herbert Tracey received his BS from Towson University and a MS from Johns Hopkins University. Currently, he is an instructor of mathematical sciences at Loyola University Maryland and served as department chair of mathematics (retired) at Hereford High School.

Jonathan Schofield graduated in civil engineering from University of Bradford in the UK, where he works "with water." He provided the essential final pass to quadruple-check the numbers in the book.

Acknowledgments

Our editors:

Thanks to **Courtney Nash**, who supported our efforts to think beyond the conventional curriciulum and took time to wonder with us about what the learner's brain wanted to know next. She consistently pushed us toward the bigger picture—and it's a much better book for that. Extra props to her for googling our odd British phrases and sayings to find U.S.-friendly translations!

Courtney Nash

Brett McLaughlin

And to **Brett McLaughlin**, who started us off on this book, and provided some really kick-ass training on the Head First way and why it rules the world. Also to the folk who rocked training in Boston 2008 and added so much to the experience: Lou Barr, Elizabeth Robson, John Guenin, Edward Ocampo-Gooding, Aaron Glimme, Dave Sussman, Les Hardin, David Flatley and Tracey Pilone. (We heard everybody named in person buys a copy....)

Our artworker and brain-fuzz-detector:

Badger

To **Badger** (Jennie Routley), for the initial "you should write one of those…I could do the pictures!", through to many hours of genius graphics work. Also for never holding back on telling us that some algebra was "disgusting" when we'd crossed the line.

The O'Reilly team:

To **Karen Shaner**, who handled the tech-review process, and provided a pep talk when the comments first started coming. To our production editor, **Rachel Monaghan**, for being patient about the fact that we mistyped "Length" eight times in one chapter. To **Lou Barr**, for her genius Head First template. And to **Scott DeLugan** and **Sanders Kleinfeld**, for once again going above and beyond to get the book out.

Lindsey's friends and family:

To **Helen**, for understanding that when I say, "I'll be there in a minute…I'm almost done…", it's an expression of hope and not reality. And for accommodating a year of lost weekends and working-on-vacation, and never, ever being intolerant of yet another conversation about triangles. And to **Joe**—my brilliant, dyspraxic stepson—for being our "learning differences" guinea pig.

Dawn's friends and family:

Work on this book would have been lot harder without my amazing support network of family and friends. Special thanks go to David, Mum and Dad, Carl, Steve, Gill, and Paul. I've truly appreciated all your support and encouragement.

And finally: To **Bert Bates** and **Kathy Sierra**, for creating the series that changed our lives.

Safari® Books Online

 Safari® Books Online is an on-demand digital library that lets you easily search over 7,500 technology and creative reference books and videos to find the answers you need quickly.

With a subscription, you can read any page and watch any video from our library online. Read books on your cell phone and mobile devices. Access new titles before they are available for print, and get exclusive access to manuscripts in development and post feedback for the authors. Copy and paste code samples, organize your favorites, download chapters, bookmark key sections, create notes, print out pages, and benefit from tons of other time-saving features.

O'Reilly Media has uploaded this book to the Safari Books Online service. To have full digital access to this book and others on similar topics from O'Reilly and other publishers, sign up for free at *http://my.safaribooksonline.com*.

1 finding missing angles

✳ *Reading between* ✳
✳ *the lines*

Ever get the feeling there's something they're not telling you?

If you want to master the **real world**, you need to get geometry. It's a set of **tools** for turning a little bit of information into a **complete picture**. Whether you want to design something, build something or find out how a situation really went down, geometry can make sure you've always got **the lowdown**. So if you want to keep in the loop, grab your hat, pack your pencil, and join us on the bus to Geometryville.

There's been a homicide

And the number one suspect, Benny, is firmly behind bars. To the officers it looked like an open and shut case but Benny is still claiming he's innocent.

So yeah, I owed Micky a little money and went out to meet him at some waterfront warehouse. Next thing I know, I hear a buncha noise, and Micky's lying dead, shot through the back! I was gonna pay him, I swear. It wasn't that much money, not enough to off the guy over it.

BENNY SMITH
HF COUNTY
SHERIFF'S OFFICE

GEO180

BULLET FOUND STUCK IN WALL HERE

CRIME SCENE · DO NOT CROSS LINE · CRIME SCENE · DO NOT CROSS LINE · CRIME SCENE · DO N

Crime scene photo

There are very few clues to go on, so the CSIs are relying on you to work up the only solid clues they've got: the ballistics evidence.

In the ballistics lab you've got to cover all the angles

As the ballistics investigator, your mission is to work out what happened to the bullet between when it was fired from the gun and when it stopped moving, and whether that ties up with what the investigators suspect—that Benny shot Micky.

Bullets travel in straight lines

The whole basis of your investigation is something that all bullets have in common: **they travel in straight lines**.

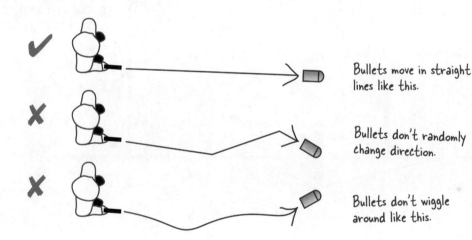

Bullets move in straight lines like this.

Bullets don't randomly change direction.

Bullets don't wiggle around like this.

Angles matter

Angles are formed where straight lines meet. Because bullets move in straight lines, *taking aim* really means deciding what the **angle** is between you and the target. The angle of the shot determines whether the bullet hits or misses the target.

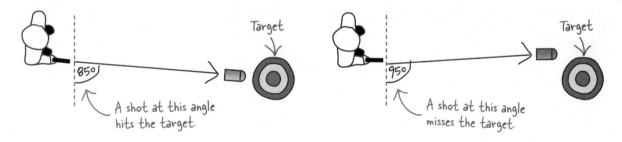

Target

85°

A shot at this angle hits the target.

Target

95°

A shot at this angle misses the target.

The case rests upon you finding the answer to just one question...

Do the angles between Benny, Micky, and the bullet match up?

At the crime scene, the investigators took some measurements and sketched the positions of Benny, Micky's body, and the point where they found the bullet in the wall, all relative to the building where the crime took place.

Top-down view of the crime scene

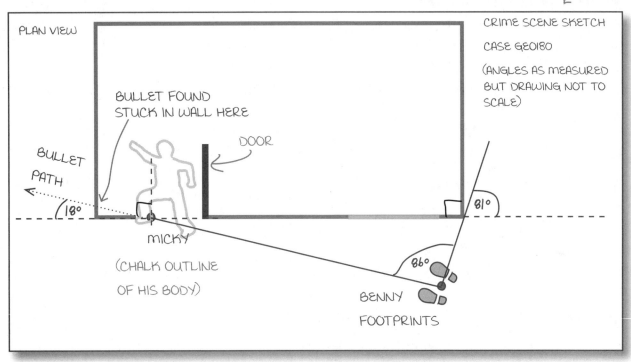

They must have been busy that day, because they didn't measure everything! But, they found one important detail: the bullet entered the wall at an angle of 18° through the front wall.

So, to solve the crime, all you have to do is prove that from where Benny was standing to the target (Micky), the bullet would have traveled in a straight line that joins up perfectly with the bullet path.

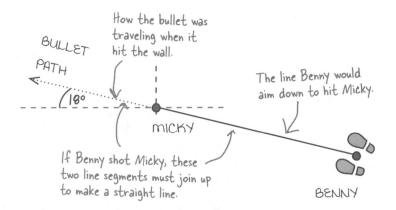

4 *Chapter 1*

Well, this is easy! You can see just by looking at it that the line from Benny to Micky and the line for the bullet path join up to make one straight line.

Jill: Hang on—I'm not sure we can really go by eye like that!

Joe: Oh, well, we could use a ruler to check it. But I'm pretty sure it's a straight line.

Frank: I don't think we could trust it, even if the ruler showed it was straight—the sketch clearly says "drawing not to scale." The angles we've been given are correct, but if you checked it with a protractor the lines on the sketch wouldn't necessarily be the same angles as at the scene itself.

Joe: What? Well, it's useless then, isn't it?

Frank: It's not useless—if it says 18° on the sketch then it was 18° at the scene because they measured it there. But I don't think we can go on what the sketch *looks* like. We're going to actually have to work out whether the line segments really join up into one straight line.

Joe: But we've only got three angles to go on. I bet the chief will say we need to fill them all in. He's gonna be mad.

Jill: What if we could find a way to guess some angles based on other angles or something? But that just sounds really inaccurate—not exactly good for our case in court!

Frank: It sounds like the right kind of approach though. And anyway, I think they've measured five angles, not just three....

⚛ BRAIN POWER

Does the sketch tell you five angles, or just three?

Is there a way you could start to find some of the angles you haven't been given on the sketch?

Right angles aren't always marked with numbers

A right angle is an internal angle (the smallest angle) between two lines equal to 90°. Instead of drawing a curve, like we do for most angles, we mark it with a square corner.

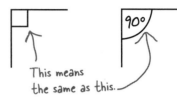

This means the same as this.

A right angle is a quarter turn—like the angle between the hands on the clock at 3 o'clock or the amount you have to turn a skateboard to get into and out of a boardslide without landing on your face.

It is still possible to land on your face even if you make the 90° turn.

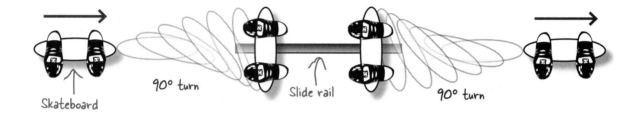

Skateboard

90° turn

Slide rail

90° turn

Lines that meet at a right angle are perpendicular

Whether they actually cross each other or only meet at a point, two lines that form a right angle are called **perpendicular** lines. We could also say that one of the lines is **perpendicular to** the other.

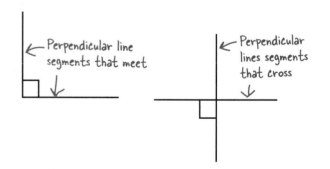

Perpendicular line segments that meet

Perpendicular lines segments that cross

So it turns out that we do have five angles on the sketch after all.

Although they don't say "90°" on them, the little square angle marks tell us that some of the angles are right angles.

But what about all those other angles that aren't marked on the sketch?

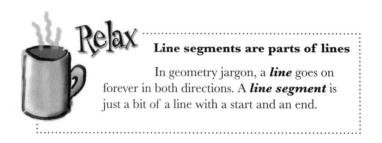

Relax

Line segments are parts of lines

In geometry jargon, a **line** goes on forever in both directions. A **line segment** is just a bit of a line with a start and an end.

Angles can be made up of other, smaller angles

If you cut up an angle into pieces, the smaller angles add up to the original angle.

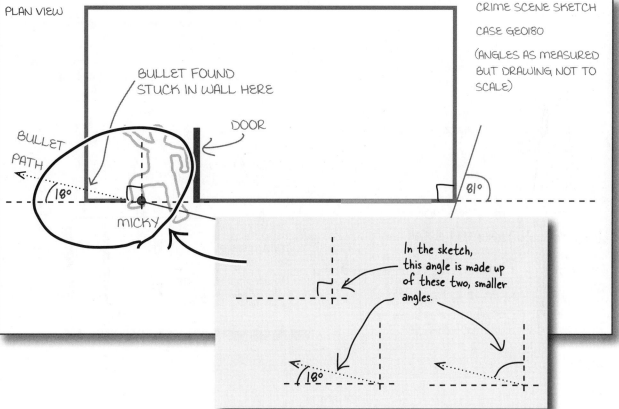

PLAN VIEW

CRIME SCENE SKETCH

CASE GEO180

(ANGLES AS MEASURED BUT DRAWING NOT TO SCALE)

BULLET FOUND STUCK IN WALL HERE

DOOR

BULLET PATH

18°

MICKY

81°

In the sketch, this angle is made up of these two, smaller angles.

18°

Angle Magnets

Which of these angles matches the mystery angle?

62°   72°   45°   67°

18°   ?°

Angle Magnets Solution

Which of these angles matches the mystery angle?

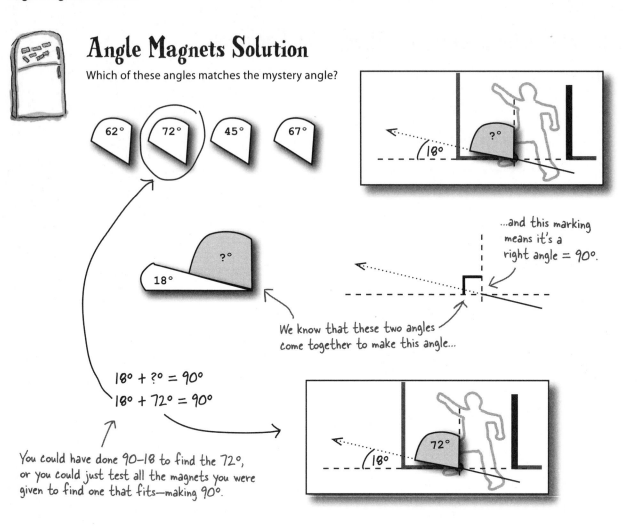

62° 72° 45° 67°

?°

18°

18° + ?° = 90°
18° + 72° = 90°

...and this marking means it's a right angle = 90°.

We know that these two angles come together to make this angle...

You could have done 90–18 to find the 72°, or you could just test all the magnets you were given to find one that fits—making 90°.

---

there are no Dumb Questions

Q: I checked that mystery angle with a protractor and it wasn't 72°—so how come you're saying it is?

A: In geometry, unless you're specifically told to measure an angle, assume that the *drawing* isn't accurate, but that the *numbers* on the sketch are. We **calculate** missing angles rather than **measure** them.

Q: What about all the other lines on the diagram? Do we know if they're straight?

A: The other line segments on the sketch represent things like walls or the path between two points. They're all definitely straight. In fact, the two line segments we're interested in—the bullet path and the path from Benny to Micky—are also straight. What we need to find out is whether they join up to form one single straight line.

Complementary angles always add up to a right angle (90°)

If two angles combine to creat a right angle we call them *complementary angles*. Usually, complementary angles are adjacent (next to each other), but they can be any two angles anywhere that add up to 90 degrees.

We say that the angles **complement** each other.

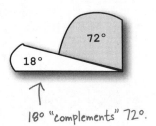

18° + 72° = 90°
so these angles are
complementary.

↑
18° "complements" 72°.

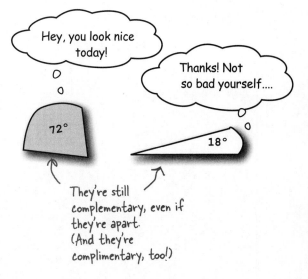

Hey, you look nice today!

Thanks! Not so bad yourself....

They're still
complementary, even if
they're apart.
(And they're
complimentary, too!)

Exercise

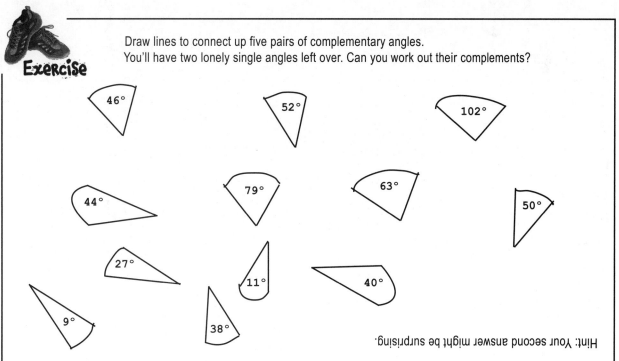

Draw lines to connect up five pairs of complementary angles.
You'll have two lonely single angles left over. Can you work out their complements?

46°   52°   102°

44°   79°   63°   50°

27°   11°   40°

9°   38°   Hint: Your second answer might be surprising.

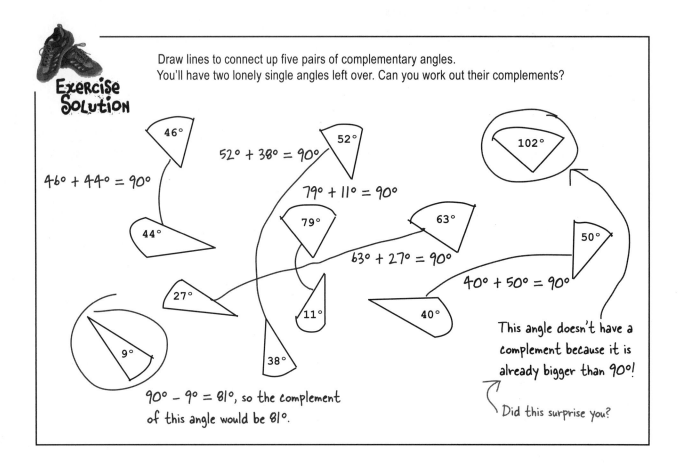

Exercise Solution

Draw lines to connect up five pairs of complementary angles.
You'll have two lonely single angles left over. Can you work out their complements?

46°

52°

102°

52° + 38° = 90°

46° + 44° = 90°

79° + 11° = 90°

79°

63°

44°

50°

63° + 27° = 90°

27°

40° + 50° = 90°

9°

11°

40°

This angle doesn't have a complement because it is already bigger than 90°!

38°

Did this surprise you?

90° − 9° = 81°, so the complement of this angle would be 81°.

there are no Dumb Questions

Q: How can I tell which angles do and don't have a complement?

A: Only **acute** angles—angles less than 90°—can be complementary. An **obtuse** angle is already greater than 90° on its own.

Q: Why doesn't 102° have a negative complement of −12°? Surely that would add up to 90°?

A: Complements can never be negative. There's no real reason for this except that the term "complementary" means two positive angles which add up to 90°.

Only acute angles can be complementary.

Right angles often come in pairs

A right angle is a quarter turn, so two right angles add up to one half turn—a 180° on your skateboard, assuming you land fakey (facing the opposite way to when you started).

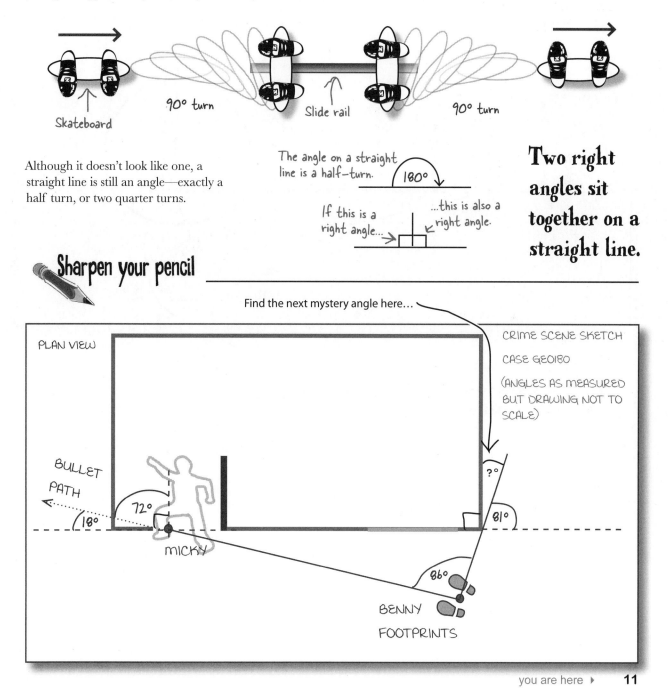

90° turn

Skateboard

Slide rail

90° turn

Although it doesn't look like one, a straight line is still an angle—exactly a half turn, or two quarter turns.

The angle on a straight line is a half-turn.

180°

If this is a right angle...

...this is also a right angle.

Two right angles sit together on a straight line.

Sharpen your pencil

Find the next mystery angle here...

PLAN VIEW

CRIME SCENE SKETCH

CASE GEO180

(ANGLES AS MEASURED BUT DRAWING NOT TO SCALE)

BULLET PATH

72°

18°

?°

81°

MICKY

86°

BENNY

FOOTPRINTS

Sharpen your pencil
Solution

Find the next mystery angle here...

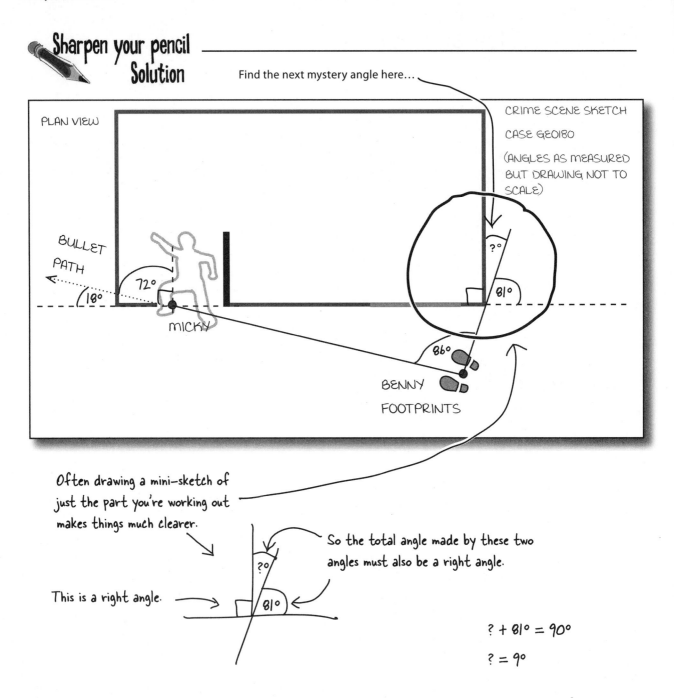

PLAN VIEW

CRIME SCENE SKETCH

CASE GEO180

(ANGLES AS MEASURED BUT DRAWING NOT TO SCALE)

BULLET PATH

72°

18°

MICKY

?°

81°

86°

BENNY

FOOTPRINTS

Often drawing a mini-sketch of just the part you're working out makes things much clearer.

So the total angle made by these two angles must also be a right angle.

This is a right angle.

?°

81°

$? + 81° = 90°$

$? = 9°$

> Great. Whatever. But what on earth has working out the angle behind the corner of the building got to do with proving that Benny shot Micky all the way over on the other side?

Often you need to solve the puzzle piece by piece.

Each individual angle on the sketch doesn't tell you much, but together they make up an accurate picture that shows you how every part fits in relation to the other parts.

It would be great to jump straight to the most important angle—if you even knew what that was—but usually we need to find a bunch of less exciting stuff to help us get there.

It's like any other puzzle solving really—you're not always fighting the end-of-level boss, sometimes you're just collecting the power-ups you need to beat him/her/it when you get there!

BRAIN BARBELL

We know that you can always fit a pair of right angles on a straight line, but what if we don't have right angles?

What might this pair of mystery angles (a and b) from our crime scene sketch add up to?

You can assume that all the lines shown here are straight.

9°

81°

a°   b°

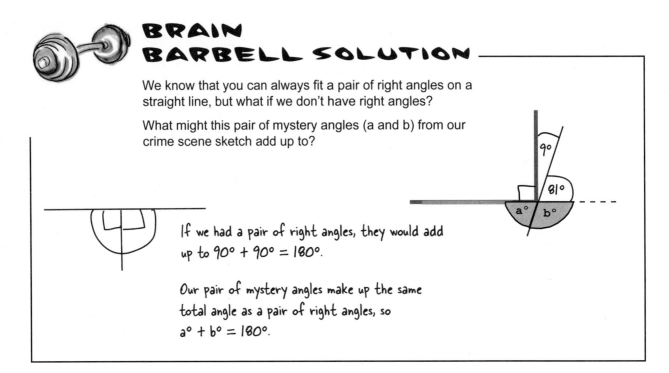

BRAIN BARBELL SOLUTION

We know that you can always fit a pair of right angles on a straight line, but what if we don't have right angles?

What might this pair of mystery angles (a and b) from our crime scene sketch add up to?

If we had a pair of right angles, they would add up to $90° + 90° = 180°$.

Our pair of mystery angles make up the same total angle as a pair of right angles, so $a° + b° = 180°$.

Angles on a straight line add up to 180°

Whether a half turn is made up of two quarter turns or lots of different small turns, the total angle of a half turn is always 180°.

This means that when the angle on a straight line is divided up into smaller angles we can always be sure that they add up to 180°.

Half a pizza is half a pizza, no matter how you slice it.

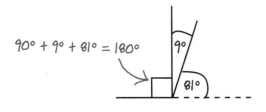

$90° + 9° + 81° = 180°$

Angle Magnets

Fit the loose angle magnets into the gaps to form three complete straight lines. Some gaps might need more than one magnet to fill them.

You might need to try a few different combinations, so don't be shy!

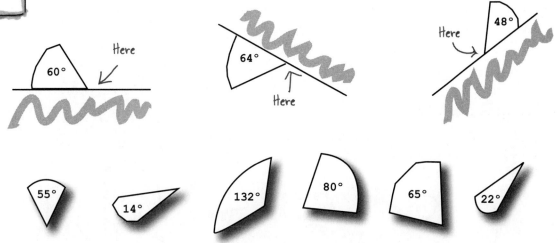

Angle Magnets Solution

Fit the loose angle magnets into the gaps to form three complete straight lines.

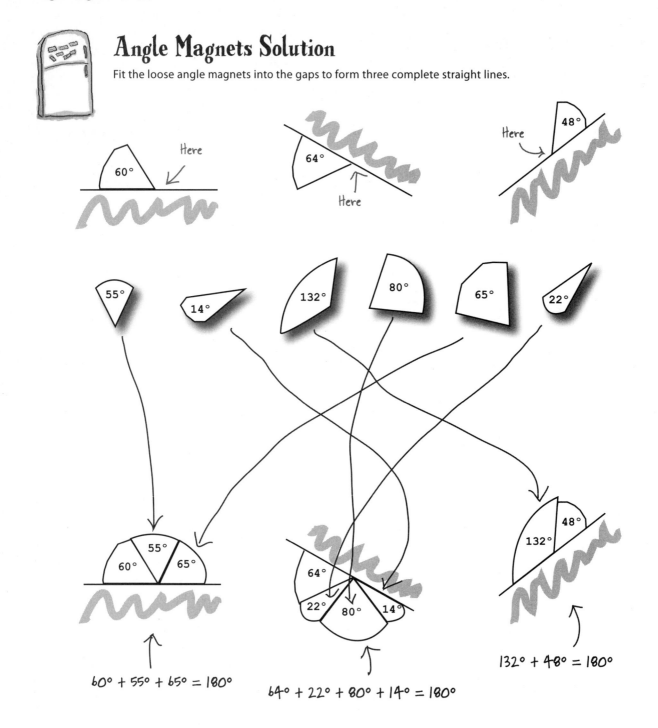

$60° + 55° + 65° = 180°$

$64° + 22° + 80° + 14° = 180°$

$132° + 48° = 180°$

<u>Pairs</u> of angles that add up to 180° are called <u>supplementary</u> angles

Any **two** angles which add up to 180° are known as
supplementary angles. They are easiest to spot when
they're on a straight line, but they can also be far apart.

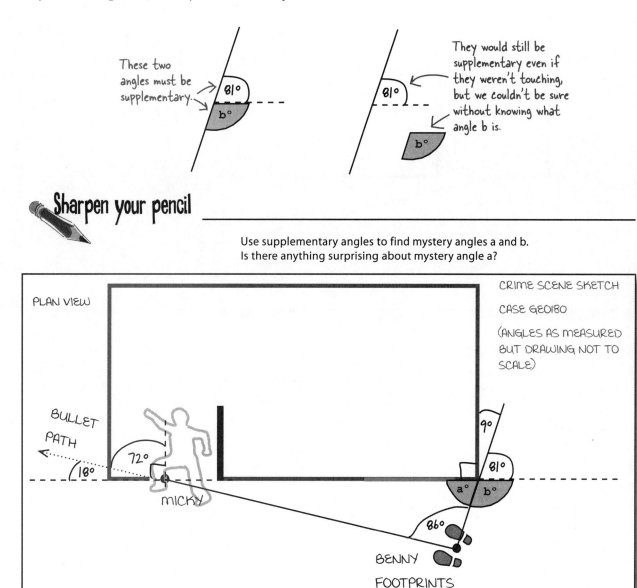

These two angles must be supplementary.

81°

b°

They would still be supplementary even if they weren't touching, but we couldn't be sure without knowing what angle b is.

81°

b°

Sharpen your pencil

Use supplementary angles to find mystery angles a and b.
Is there anything surprising about mystery angle a?

PLAN VIEW

CRIME SCENE SKETCH

CASE GEO180

(ANGLES AS MEASURED BUT DRAWING NOT TO SCALE)

BULLET PATH

72°

18°

MICKY

9°

81°

a° b°

86°

BENNY

FOOTPRINTS

Sharpen your pencil
Solution

Use supplementary angles to find mystery angles a and b.
Is there anything surprising about mystery angle a?

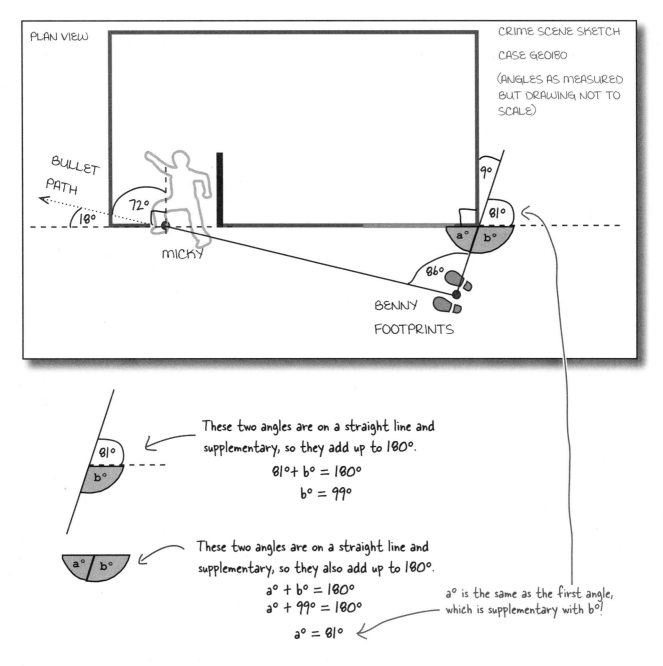

PLAN VIEW

CRIME SCENE SKETCH

CASE GEO180

(ANGLES AS MEASURED BUT DRAWING NOT TO SCALE)

BULLET PATH

72°

18°

MICKY

9°

81°

a°   b°

86°

BENNY

FOOTPRINTS

81°

b°

These two angles are on a straight line and supplementary, so they add up to 180°.

$$81° + b° = 180°$$
$$b° = 99°$$

a°   b°

These two angles are on a straight line and supplementary, so they also add up to 180°.

$$a° + b° = 180°$$
$$a° + 99° = 180°$$

$$a° = 81°$$

a° is the same as the first angle, which is supplementary with b°!

Vertical angles are always equal

When two straight lines cross they always create two pairs of equal opposite angles, called **vertical angles**.

Each angle has to form a supplementary pair with either of the angles on the other side of it, so they must be equal.

Save yourself a ton of math by spotting equal pairs of vertical angles.

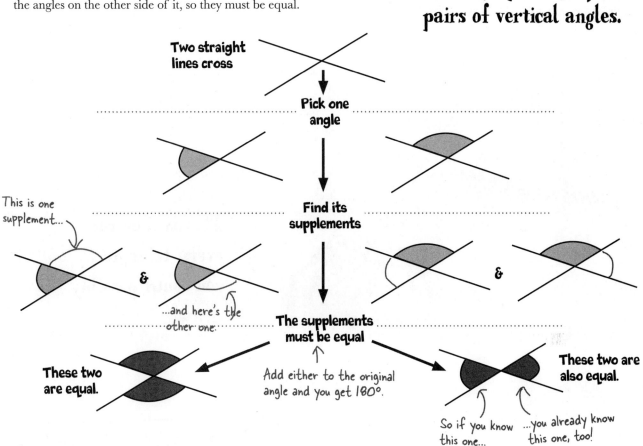

Two straight
lines cross

Pick one
angle

Find its
supplements

This is one
supplement...

&

...and here's the
other one.

The supplements
must be equal

Add either to the original
angle and you get 180°.

These two
are equal.

&

These two are
also equal.

So if you know
this one...

...you already know
this one, too!

Patterns in geometry aren't just coincidence

These **relationships** are how geometry works...and to see a really freaky one, try this:

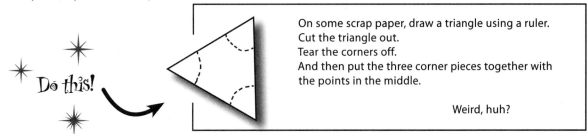

Do this!

On some scrap paper, draw a triangle using a ruler.
Cut the triangle out.
Tear the corners off.
And then put the three corner pieces together with the points in the middle.

Weird, huh?

The corner angles of a triangle always add up to a straight line

The corner angles of a triangle always make a straight line, no matter what kind of triangle it is. Angles on a straight line always add up to 180°, so the corner angles of a triangle must always add up to 180°, too.

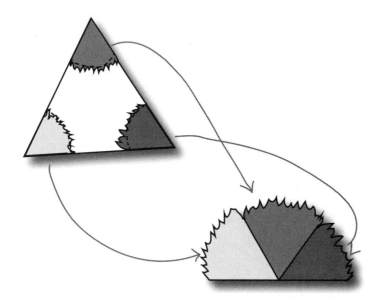

This is true for every triangle you could possibly draw!

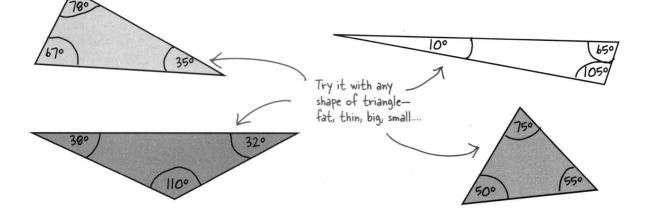

Try it with any shape of triangle— fat, thin, big, small....

Find one more angle to crack the case

There's only one unknown angle left on the diagram. Once you've figured out this one you can charge Benny with the shooting for sure.

Exercise

Find this last mystery angle to complete your investigation and prove that Benny was the shooter.

How does it fit in with what you've already found out?

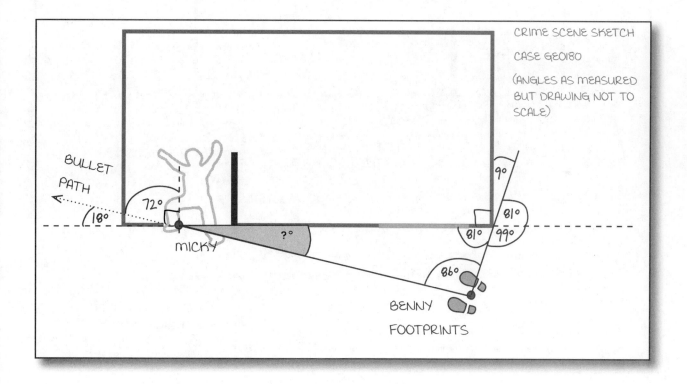

Exercise Solution

Find this last mystery angle to complete your investigation and prove that Benny was the shooter.

How does it fit in with what you've already found out?

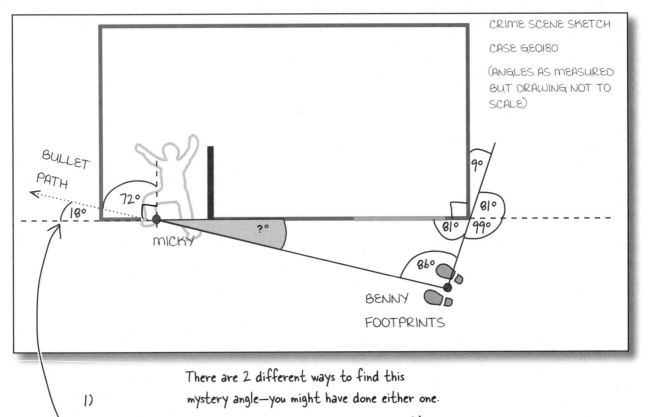

There are 2 different ways to find this mystery angle—you might have done either one.

1)

Angle c is opposite this angle here
And vertical angles are equal, so ? = 18°

But this doesn't fit with the triangle angles:

18° + 81° + 86° = 185°

They should add up to 180°.

Hmm...something funny is going on. It doesn't matter how you worded it, as long as you noticed something didn't add up.

2)

Angles in a triangle add up to 180°
so 81° + 86° + ? = 180°
? = 180° – 81° – 86° = 13°

But this doesn't fit with the 18° angle it's opposite. They should be equal.

Something doesn't add up!

Using the two different methods for finding the mystery angles gives us two different results.

And vertically opposite angles are equal, so ? = 18°

Angles in a triangle add up to 180° so 81° + 86° + ? = 180°
? = 180° − 81° − 86° = 13°

This is not good. The whole investigation could be compromised by bad math!

Joe: This is a disaster. How can there be two different answers? It's totally confusing.

Jill: Benny's lawyer is gonna have a field day. I don't believe it. We've got the right guy in custody—it's obvious he did it—and we're gonna have to let him go because of some messed up math! Can we fix it?

Frank: The weirdest thing is that I'm looking again at the math for either method and it looks good. I mean—corners on a triangle **always** add up to 180 degrees, and vertically opposite angles are **always** equal. Those are the rules.

Joe: I knew we were relying too much on these coincidences. My guess is that these "rules" don't always actually work!

Jill: I really don't think that's what it means. If we haven't messed up the calculations then the fact that we got two different answers simply has to mean something else....

What do you think it means?

Frank   Jill   Joe

If it doesn't all add up, then something isn't as it seems

The mystery angle is 18° **and** the mystery angle is 13°…well, that clearly doesn't add up. If the mystery angle is 13°, then it can't be vertically opposite our 18° bullet path angle as well. This means that the line on the sketch from Benny to Micky doesn't join up perfectly with the bullet path.

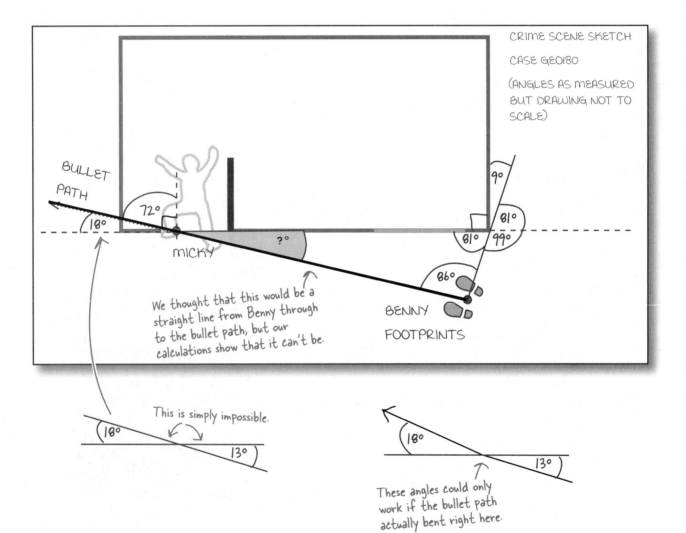

CRIME SCENE SKETCH

CASE GEO180

(ANGLES AS MEASURED BUT DRAWING NOT TO SCALE)

BULLET PATH

18°

72°

MICKY

?°

We thought that this would be a straight line from Benny through to the bullet path, but our calculations show that it can't be.

9°

81°

81°   99°

86°

BENNY FOOTPRINTS

This is simply impossible.

18°

13°

18°

13°

These angles could only work if the bullet path actually bent right here.

You've proved that Benny couldn't have shot Micky!

Unless Benny has a magic gun, he couldn't have shot Micky AND had the bullet enter the wall at the point that the CSIs found at the scene.

Benny is not our shooter, so the charges have been dropped and he's been released from custody, which has produced an interesting development....

> What I tried to tell them when I was arrested is that I don't even own a gun! But there was this other guy there, Charlie, inside the building. I think he took a shot at me. I heard a gunshot and breaking glass so I ducked. The bullet hit the car that was parked behind me, and then it sped away. I don't know how Micky ended up shot, but at least you know it wasn't me!

And the chief is on the phone...

> Great work. This case was a mess before you started working up the ballistics. But who the heck is Charlie? We'll get the team back on the scene to search for evidence. Stand by—we're gonna need your help.

Benny, seriously relieved to be back in his own wardrobe choices

We've got a new sketch—now for a new ballistics report

Following up on Benny's story, officers have returned to the scene, and found footprints, a broken window, and a set of tire tracks that might indicate that he's telling the truth. But there's a big problem—from where Charlie appeared to be standing, he can't have shot Micky through the door, plus the bullet that killed him came from the outside the building!

We do know that the bullet must have traveled in a straight line headed in the direction shown on the sketch, but how it got there is the big question.

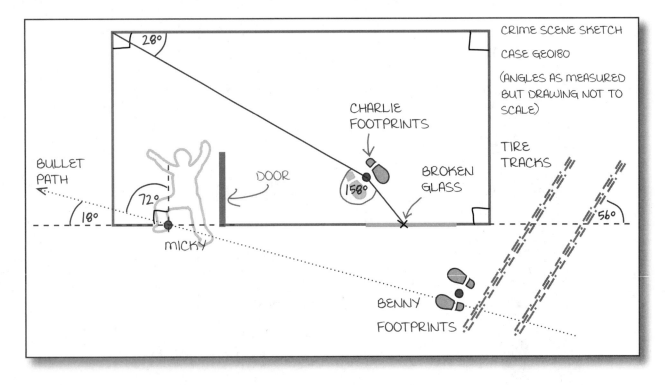

there are no
Dumb Questions

Q: **The dotted line extending the bullet path—how do we know it's really straight?**

A: We're going to start by saying that this line MUST be straight heading backward from where the bullet entered the wall, and then work up the sketch to find out how the bullet came to be traveling in exactly that trajectory.

We need a new theory

Although you don't **know** what happened, if you can come up with some imaginative suggestions about what **might** have happened, you can test them by working out whether the angles add up, just like when you proved that Benny couldn't have shot Micky.

> Right. So now's the time to start guessing? Maybe some aliens came down and shot Micky? This is stupid.

It's not just a matter of guessing.

You can use the statement Benny gave and all the relevant information on the crime scene sketch to guide you. Our points, lines, and angles are still as true as ever.

Statement by Benny on release from custody

What I tried to tell them when I was arrested is that I don't even own a gun! But there was this other guy there, Charlie, inside the building. I think he took a shot at me. I heard a gunshot and breaking glass so I ducked. The bullet hit the car that was parked behind me—it sped away. I don't know how Micky ended up shot, but at least you know it wasn't me!

Sharpen your pencil

Read over Benny's statement. Using the sketch on the left page, or a blank sheet of paper, can you come up with a new idea about how Micky got shot?

(You can assume it didn't involve aliens.)

Sharpen your pencil
Solution

Read over Benny's statement. Using the sketch on the left page, or a blank sheet of paper, can you come up with a new idea about how Micky got shot?

A new theory: Charlie shot at Benny, breaking the glass in the window, but missed, hit a parked car, and the bullet bounced off and hit Micky

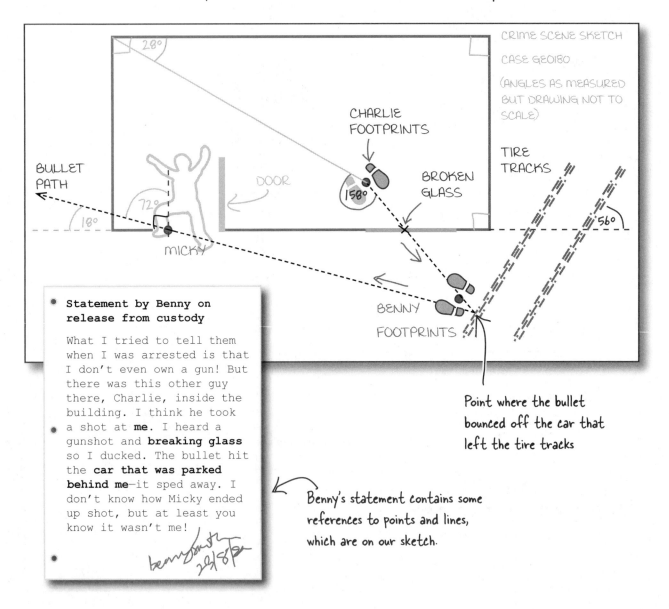

CRIME SCENE SKETCH
CASE GEO180
(ANGLES AS MEASURED BUT DRAWING NOT TO SCALE)

28°

CHARLIE FOOTPRINTS

BROKEN GLASS

TIRE TRACKS

BULLET PATH

DOOR

158°

72°

18°

56°

MICKY

BENNY FOOTPRINTS

Statement by Benny on release from custody

What I tried to tell them when I was arrested is that I don't even own a gun! But there was this other guy there, Charlie, inside the building. I think he took a shot at **me**. I heard a gunshot and **breaking glass** so I ducked. The bullet hit the **car that was parked behind me**—it sped away. I don't know how Micky ended up shot, but at least you know it wasn't me!

Point where the bullet bounced off the car that left the tire tracks

Benny's statement contains some references to points and lines, which are on our sketch.

Work out what you need to know

Based on your sketch, you need to work out what your theory actually means in terms of points, lines, and angles.

In this case we've got a few important details. We have a line from the point where Charlie was standing, through the hole in the window, until it hits the parked car. We also sketched a line moving backward from the bullet, until that hits the parked car, too.

If our theory is true, and Charlie was the shooter, the way the bullet bounced off the car is important.

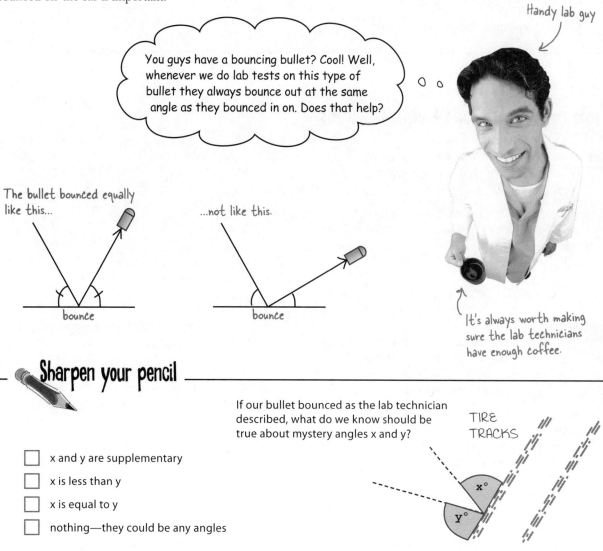

Handy lab guy

You guys have a bouncing bullet? Cool! Well, whenever we do lab tests on this type of bullet they always bounce out at the same angle as they bounced in on. Does that help?

It's always worth making sure the lab technicians have enough coffee.

The bullet bounced equally like this...

...not like this.

bounce

bounce

Sharpen your pencil

If our bullet bounced as the lab technician described, what do we know should be true about mystery angles x and y?

TIRE TRACKS

☐ x and y are supplementary

☐ x is less than y

☐ x is equal to y

☐ nothing—they could be any angles

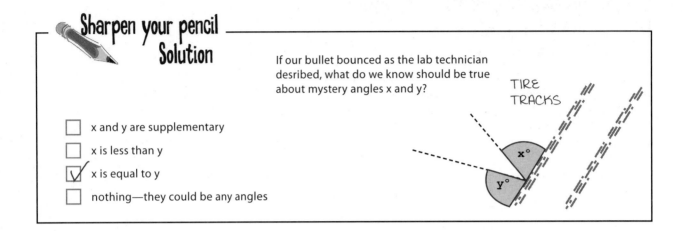

Sharpen your pencil Solution

If our bullet bounced as the lab technician desribed, what do we know should be true about mystery angles x and y?

☐ x and y are supplementary

☐ x is less than y

☑ x is equal to y

☐ nothing—they could be any angles

TIRE TRACKS

Tick marks indicate equal angles

Tick marks on angles are used to show that the angles are equal, even if we don't know what size the angles are.

Each set of equal angles has a different number of ticks, so you can mark more than one set on the same sketch if you need to.

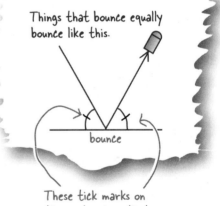

Things that bounce equally bounce like this.

bounce

These tick marks on the angles show that they are equal.

Watch it!

Remember—your sketch isn't usually accurate enough to see by eye whether two angles are equal.

Unless you're told otherwise, a sketch is just a sketch. Even if the angles drawn look pretty close to the real angles, you can't rely on your eye or a protractor—you've gotta work the angles out.

Use what you know to find what you don't know

By spotting patterns around the angles that you do know, you can work out the angles that you don't yet know. Sometimes, like with vertical angles, you don't even have to do any math.

Equal angles

Here are some patterns for you to look out for:

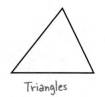

Triangles

Angles on a straight line

Complementary angles

Vertical angles

Sharpen your pencil

On the sketch, mark all the new angles you need to find to solve the crime (there are at least 11). ← These aren't the angles you found before.

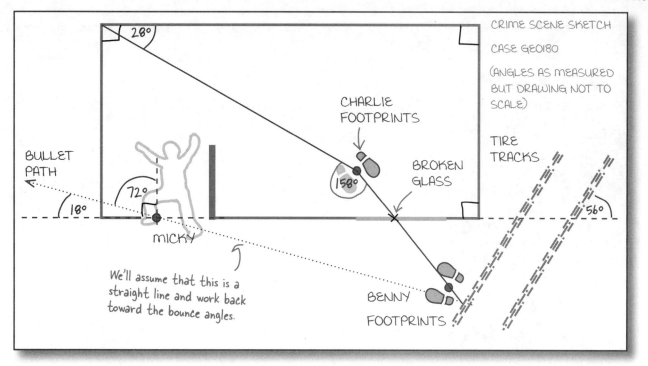

28°

CRIME SCENE SKETCH

CASE GEO180

(ANGLES AS MEASURED BUT DRAWING NOT TO SCALE)

TIRE TRACKS

CHARLIE FOOTPRINTS

BROKEN GLASS

BULLET PATH

72°

158°

18°

56°

MICKY

We'll assume that this is a straight line and work back toward the bounce angles.

BENNY FOOTPRINTS

Sharpen your pencil Solution

Mark on the sketch all the new angles you need to find to solve the crime (there are at least 11).

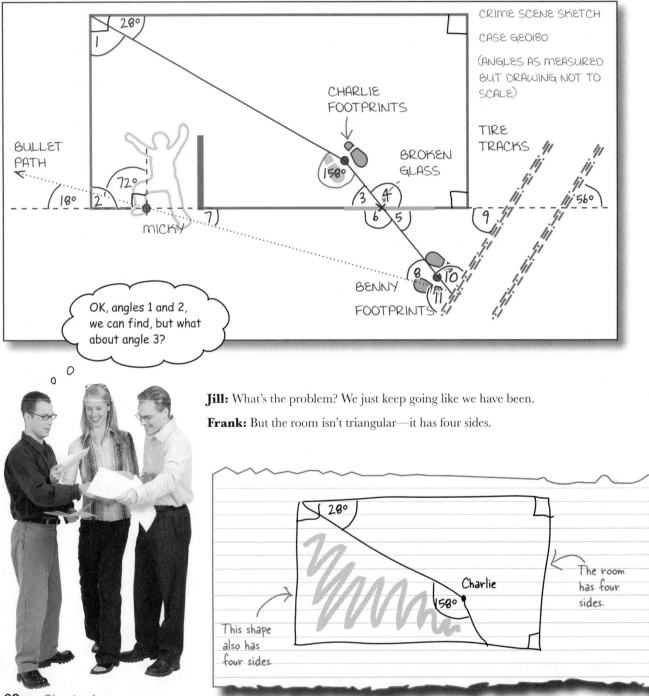

Jill: What's the problem? We just keep going like we have been.

Frank: But the room isn't triangular—it has four sides.

Jill: Oh, good point. Could we add a line of our own and chop the room into two triangles like this?

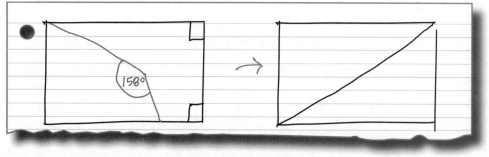

Joe: How on earth does that help?

Jill: Well…at least we know that angles in a triangle add up to 180 degrees. So if we make triangles out of the room, we can keep going like we have been. In fact, there's another shape with four sides there that could help us with angle 3—we could divide that into triangles as well maybe? Like this?

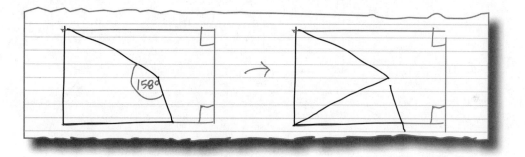

Joe: OK, but that seems like a lot more work. Maybe we could try that thing we did with the paper triangle? With the corners? That might show us something about four-sided shapes in general, and we can use that to find those missing angles?

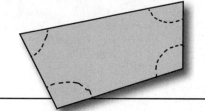

Sharpen your pencil

What is it that you need to find out about four-sided shapes in order to find the missing angles?

See whether you can figure it out, either by adding up the angles using two triangles, or tearing the corners from a four-sided shape to see what they add up to.

Sharpen your pencil
Solution

What is it that you need to find out about four-sided shapes in order to find the missing angles?

See whether either method can help you figure it out.

To find the missing angles, we need to know what angles in a four-sided shape add up to.

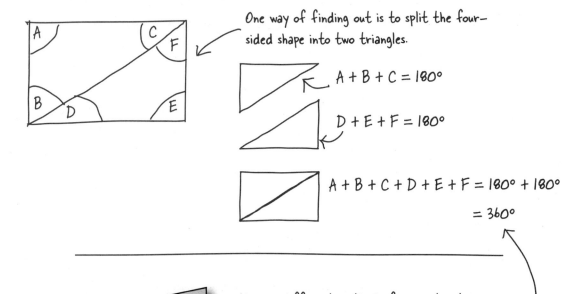

One way of finding out is to split the four-sided shape into two triangles.

$A + B + C = 180°$

$D + E + F = 180°$

$A + B + C + D + E + F = 180° + 180°$
$= 360°$

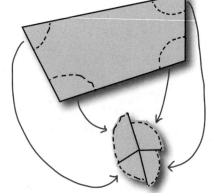

Using a different method, if you cut out a four-sided shape and tear the corners off, and put them together with the points in the middle, you'll find they make a whole turn.

And a whole turn (two half turns) is 360°.

Both methods give us the same answer: 360°.

The angles of a four-sided shape add up to 360°

Squares and rectangles have four equal angles, a quarter turn each, and even the most uneven four-sided shape still has four angles adding up to 360°.

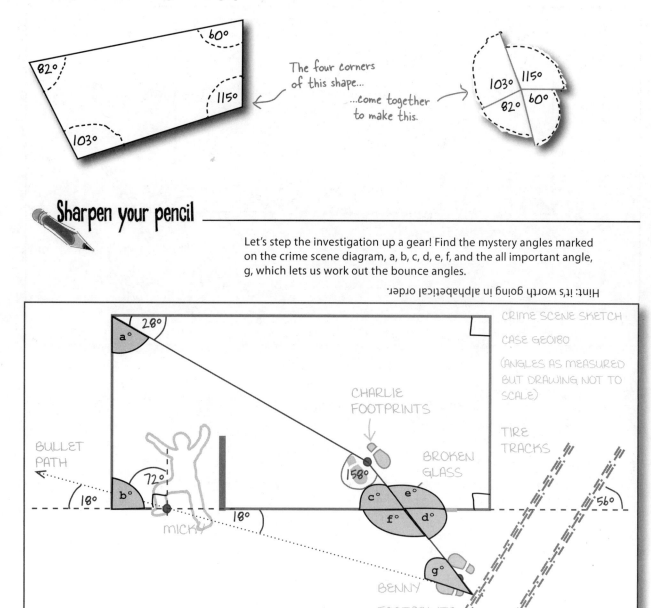

82°   60°   115°   103°

The four corners of this shape...

...come together to make this.

103° 115°
82°   60°

Sharpen your pencil

Let's step the investigation up a gear! Find the mystery angles marked on the crime scene diagram, a, b, c, d, e, f, and the all important angle, g, which lets us work out the bounce angles.

Hint: it's worth going in alphabetical order.

28°
a°

CRIME SCENE SKETCH
CASE GEO180
(ANGLES AS MEASURED BUT DRAWING NOT TO SCALE)

CHARLIE FOOTPRINTS

TIRE TRACKS

BULLET PATH

72°
b°
18°

158°

BROKEN GLASS

c° e°
f° d°

56°

18°

MICK?

g°

BENNY FOOTPRINTS

Sharpen your pencil Solution

Let's step the investigation up a gear! Find the mystery angles marked on the crime scene diagram, a, b, c, d, e, f, and the all important angle, g, which lets us work out the bounce angles.

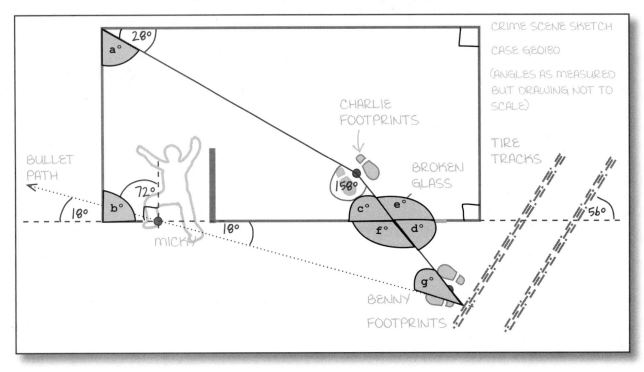

CRIME SCENE SKETCH

CASE GEO180

(ANGLES AS MEASURED BUT DRAWING NOT TO SCALE)

TIRE TRACKS

CHARLIE FOOTPRINTS

BROKEN GLASS

BULLET PATH

BENNY FOOTPRINTS

a)

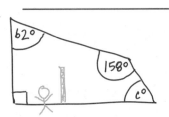

a and 28° are complementary.
a + 28° = 90°, so a = 62°

b)   The room is a four-sided shape and the other three corners are right angles, so
b = 360° − (90° + 90° + 90°) = 90°

c)   c is part of a four-sided shape, with a°
and the corner angle b° we just found, so
c = 360° − (158° + 62° + 90°) = 50°

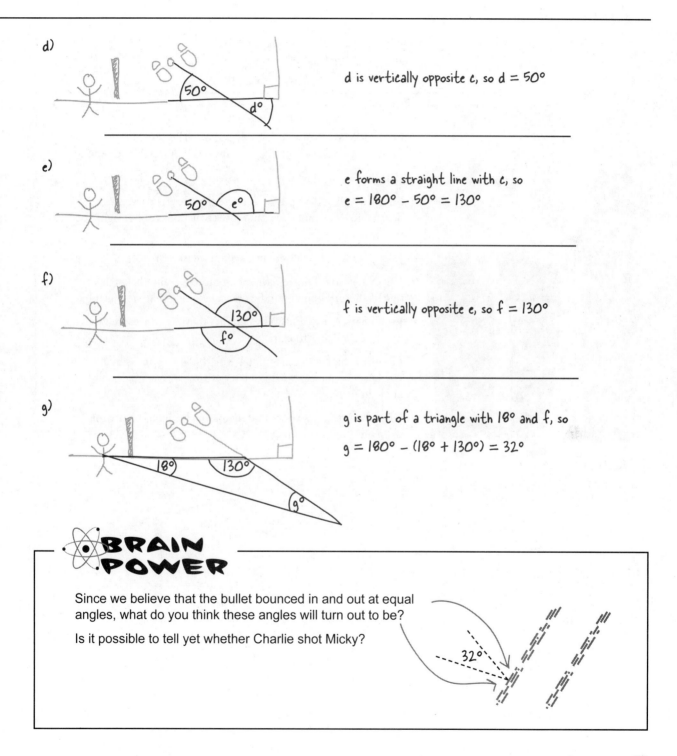

d)

d is vertically opposite c, so d = 50°

e)

e forms a straight line with c, so
e = 180° − 50° = 130°

f)

f is vertically opposite e, so f = 130°

g)

g is part of a triangle with 18° and f, so
g = 180° − (18° + 130°) = 32°

BRAIN POWER

Since we believe that the bullet bounced in and out at equal angles, what do you think these angles will turn out to be?

Is it possible to tell yet whether Charlie shot Micky?

32°

> Great! We know these two angles should be equal so we just do 180 minus 32...and then divide by 2 to get 74 degrees!

Frank: Whoa. Not so quick—based on ballistics, that's what the angles *should* be, but that's what we're still trying to establish.

Jill: But how can we check it?

Frank: Well, it depends what angle the car was parked at, doesn't it? Remember, we can't just say it "looks about right" on the sketch.

Joe: But the car isn't even marked on the sketch.

Jill: No, it isn't. But we have the tire tracks—and the angle of one of those. Though not the one on the side that the bullet hit.

Joe: That's just typical! Why couldn't they have measured the angle of the other track?

Frank: I think it doesn't matter which side they measured... I've never seen a car where one set of tires was at a different angle to the other side!

Jill: That's true...so does that mean that both sets of tire marks must be at the same angle?

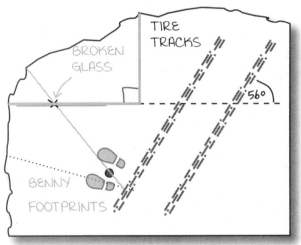

Parallel lines are lines at exactly the same angle

Two, or more, lines which are at exactly the same angle (like the tire tracks from the car must be) are called parallel lines. We use little v-shaped tick marks to indicate sets of parallel lines or line segments on a sketch.

Parallel lines can never meet or cross each other, even if you stretch them for miles and miles and miles....

Even if you stretch them forever, parallel lines never cross each other.

The distance between the lines is constant—they never get closer or farther away.

If you have more than one set of parallel lines on the same sketch then you need to use a different number of tick marks on each set to be able to tell them apart.

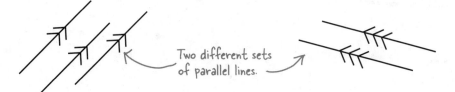

Two different sets of parallel lines.

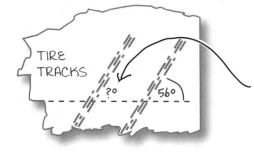

TIRE TRACKS

?°   56°

If parallel lines are always at the same angle, and our tire tracks are parallel, what should this angle here be?

Parallel lines often come with helpful angle shortcuts

When you have a line which meets or crosses a set of parallel lines, all the sets of (opposite) vertical angles that are created are the same.

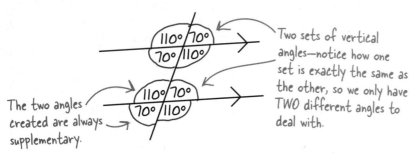

The two angles created are always supplementary.

Two sets of vertical angles—notice how one set is exactly the same as the other, so we only have TWO different angles to deal with.

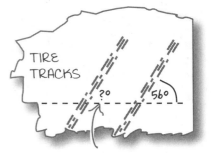

TIRE TRACKS

So this angle here should be the same—56°.

This means that you can find more missing angles without doing any math at all—just by recognizing a few patterns. The **F Pattern,** the **Z Pattern,** and the **N Pattern** are formed when parallel lines cross or meet another line.

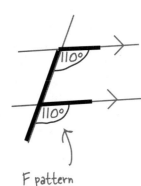

F pattern

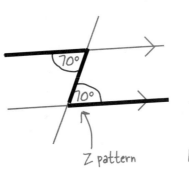

Z pattern

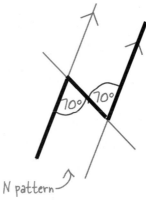

N pattern

Rotate the page and you'll see that the Z pattern and the N pattern are pretty much the same thing.

Look out for F, Z, and N patterns around parallel lines for a shortcut to finding missing angles.

Watch it!

The F, Z, or N can be hidden

Sometimes they're upside down, back-to-front, or both!

Sharpen your pencil

It's the moment of truth.

Find the crucial mystery angles, x and y, and prove once and for all whether Charlie is our shooter. If they're equal he's guilty. If they're not then the case is at another dead end.

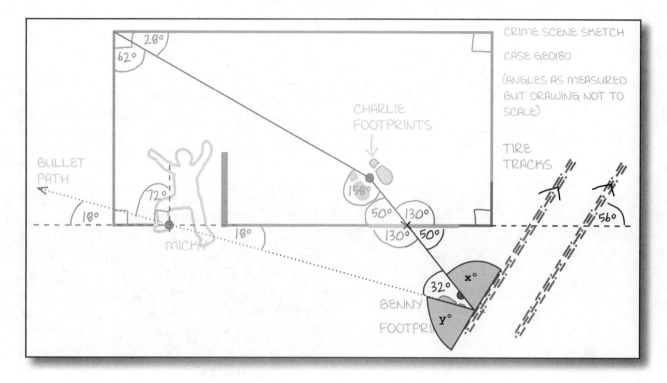

CRIME SCENE SKETCH

CASE GEO180

(ANGLES AS MEASURED BUT DRAWING NOT TO SCALE)

TIRE TRACKS

CHARLIE FOOTPRINTS

BULLET PATH

28°
62°
72°
18°
MICKY
18°
158°
50° 130°
130° 50°
56°
32°
x°
BENNY FOOTPRINTS
y°

Sharpen your pencil
Solution

It's the moment of truth.

Find the crucial mystery angles, x and y, and prove once and for all whether Charlie is our shooter. If they're equal he's guilty. If they're not then the case is at a dead end.

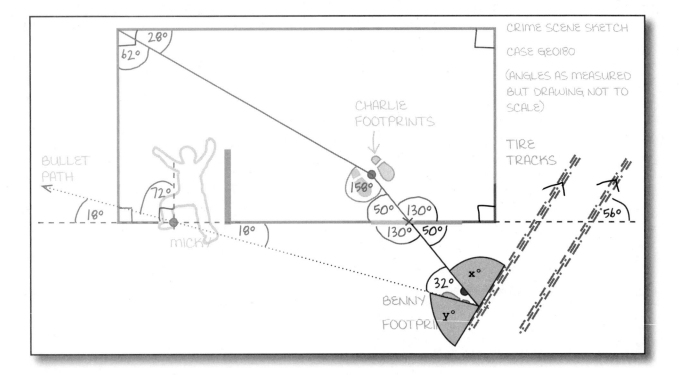

If Charlie shot Micky, then we should find that angles x and y are EQUAL, because that's how the bullet fired from his gun should bounce.

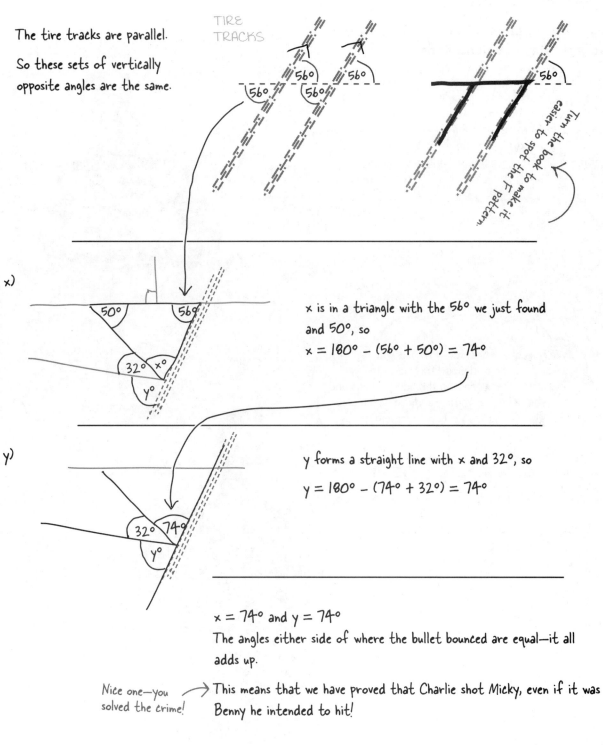

The tire tracks are parallel.

So these sets of vertically opposite angles are the same.

TIRE TRACKS

56° 56° 56° 56° 56° 56°

Turn the book to spot the T pattern. Turn the book to make it easier to spot the T pattern.

x)

50° 56°

32° x°

y°

x is in a triangle with the 56° we just found and 50°, so
x = 180° − (56° + 50°) = 74°

y)

32° 74°

y°

y forms a straight line with x and 32°, so

y = 180° − (74° + 32°) = 74°

x = 74° and y = 74°
The angles either side of where the bullet bounced are equal—it all adds up.

Nice one—you solved the crime! → This means that we have proved that Charlie shot Micky, even if it was Benny he intended to hit!

Great work—you cracked the case!

Thanks to your work uncovering the angles, the right guy is behind bars.

You've mastered a whole bunch of techniques for finding missing angles, uncovered some "sketchy" assumptions, and discovered general rules that the CSI team can use again and again to test out ballistics evidence. Your hard work and talent hasn't gone unnoticed—time to hang up your lab coat and take a hot promotion!

Charlie is gonna be making a lot of license plates....

CHARLIE GREEN
HF COUNTY
SHERIFF'S OFFICE

GEO180

We're really impressed with how you unravelled the evidence—this was a tricky one! You've been great in the ballistics lab and now you've proved you're ready to become our lead CSI, working your own important cases.

The chief couldn't be happier with you.

Before you collect your bonus and your new CSI badge, the chief would appreciate it if you left the ballistics lab a cheat sheet about how you worked out who shot Micky.

WHO DOES WHAT?

Match each technique for finding missing angles to a sketch. You can use each technique just once, but some sketches will need more than one to find the missing angle.

Angles in a triangle add up to 180°

Vertical angles are equal

Parallel lines cross other lines at the same angles

Angles on a straight line add up to 180°

Angles in a four-sided shape add up to 360°

Angles in a right angle add up to 90°

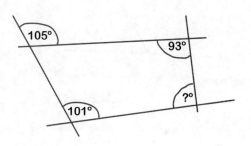

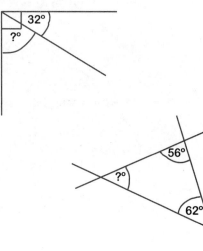

Answers on page 48.

Your Geometry Toolbox

You've got Chapter 1 under your belt, and now you've added techniques for finding missing angles to your toolbox. For a complete list of tool tips in the book, visit www.headfirstlabs.com/geometry.

Angles on a straight line add up to 180°.

Angles can be made up of other, smaller angles.

Angles in a four-sided shape add up to 360°.

Angles in a triangle add up to 180°.

A square mark indicates a RIGHT ANGLE (90°).

If two angles add up to 180°, they are SUPPLEMENTARY.

If two angles add up to 90°, they are COMPLEMENTARY.

Vertical angles are equal (and opposite each other).

Parallel lines make repeat sets of vertical angles.

F, Z, and N patterns around parallel lines save you a ton of math.

WHO DOES WHAT? SOLUTION

Match each technique for finding missing angles to a sketch. You can use each technique just once, but some sketches will need more than one to find the missing angle.

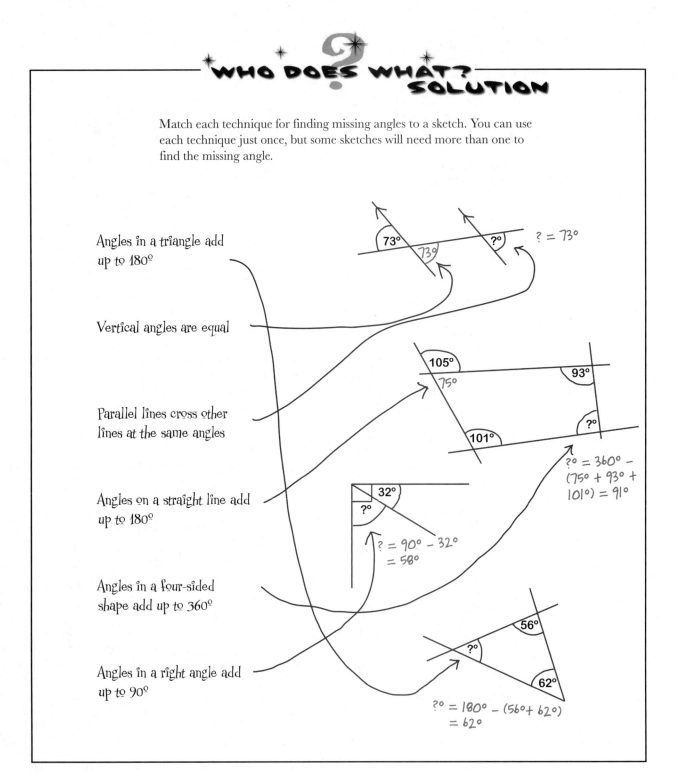

Angles in a triangle add up to 180°

Vertical angles are equal

Parallel lines cross other lines at the same angles

Angles on a straight line add up to 180°

Angles in a four-sided shape add up to 360°

Angles in a right angle add up to 90°

73°
73°
?°
? = 73°

105°
75°
93°
101°
?°
?° = 360° − (75° + 93° + 101°) = 91°

32°
?°
? = 90° − 32° = 58°

56°
?°
62°
?° = 180° − (56° + 62°) = 62°

2 similarity and congruence

Shrink to fit

He said we were very similar, but I think really he meant congruent!

In your dreams. I'm a half-inch taller than you, shorty.

Sometimes, size does matter.

Ever drawn or built something and then found out it's the **wrong size**? Or made something just perfect and wanted to *recreate it exactly?* You need *Similarity* and *Congruence*: the **time-saving techniques for duplicating** your designs smaller, bigger, or exactly the same size. Nobody likes doing the same work over—and with similarity and congruence, you'll *never have to repeat an angle calculation again.*

Welcome to myPod! You're hired

Congratulations! You've landed a dream summer job at myPod, laser etching custom stuff onto iPods, phones, and laptops. If you do well, you'll get your bonus in cool gear.

Bonus gear is up for grabs.

Class of 2010

All you have to do is prepare text and designs to be etched onto people's phones, iPods, laptops, and stuff.

Sounds cool! Think you can etch my new cell phone?

Meet Liz, your first customer.

Liz wants you to etch her phone

Time to get to work. Liz has picked a design she loves for her
new cell phone, now all you have to do is engrave it on the back.

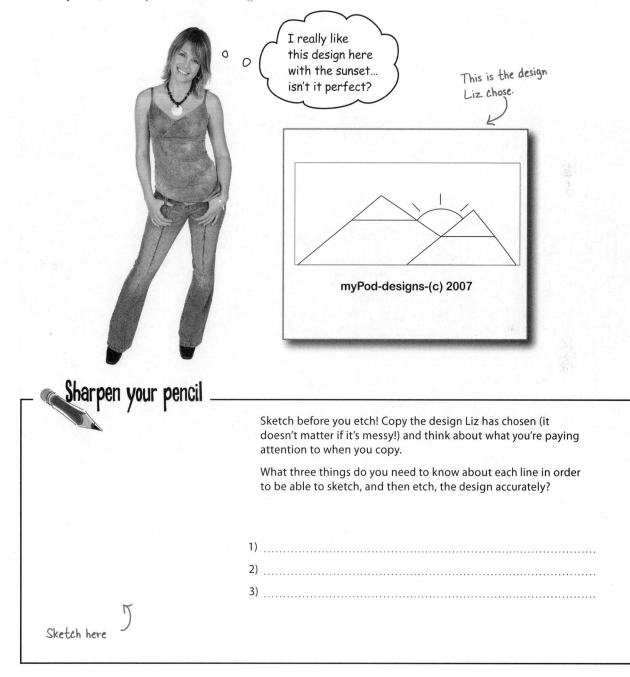

I really like this design here with the sunset... isn't it perfect?

This is the design Liz chose.

myPod-designs-(c) 2007

Sharpen your pencil

Sketch before you etch! Copy the design Liz has chosen (it
doesn't matter if it's messy!) and think about what you're paying
attention to when you copy.

What three things do you need to know about each line in order
to be able to sketch, and then etch, the design accurately?

1) ....

2) ....

3) ....

Sketch here

Sharpen your pencil
Solution

Sketch before you etch! Copy the design Liz has chosen (it doesn't matter if it's messy!) and think about what you're paying attention to when you copy.

What three things do you need to know about each line in order to be able to sketch, and etch, the design accurately?

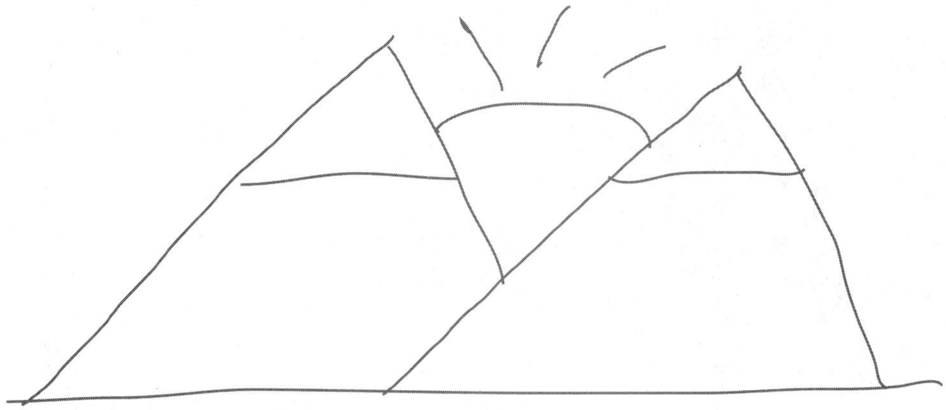

When you sketch the design, you need to think about these things:

1) How long each line is

2) What angle each line is at

3) Where the line is (what position it starts/ends)

The designer noted ~~all~~ some of the details

The designer wrote a bunch of notes around his drawing. He seems to have filled out all the line lengths, but only included a few of the angles?!?

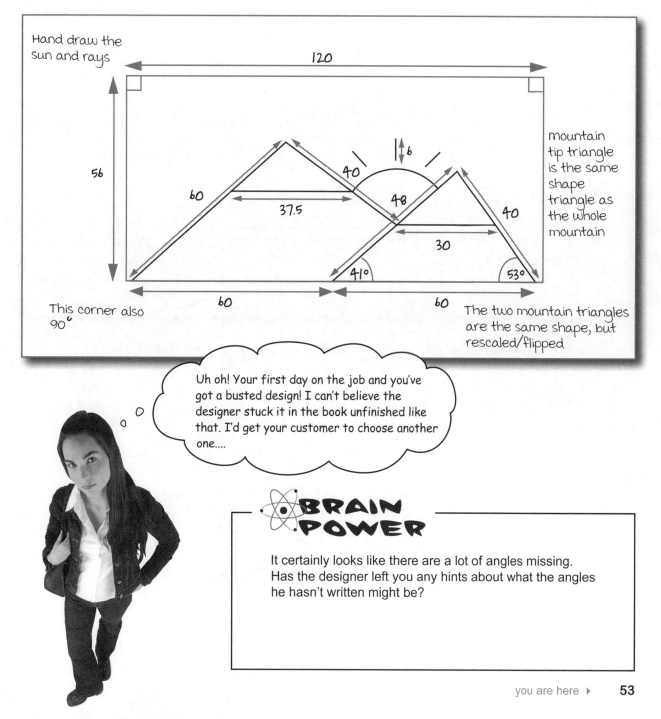

Hand draw the sun and rays

120

56

60

37.5

40

48

6

30

40

40

41°

53°

60

60

This corner also 90°

mountain tip triangle is the same shape triangle as the whole mountain

The two mountain triangles are the same shape, but rescaled/flipped

Uh oh! Your first day on the job and you've got a busted design! I can't believe the designer stuck it in the book unfinished like that. I'd get your customer to choose another one....

⚛ BRAIN POWER

It certainly looks like there are a lot of angles missing. Has the designer left you any hints about what the angles he hasn't written might be?

The design tells us that some triangles are repeated

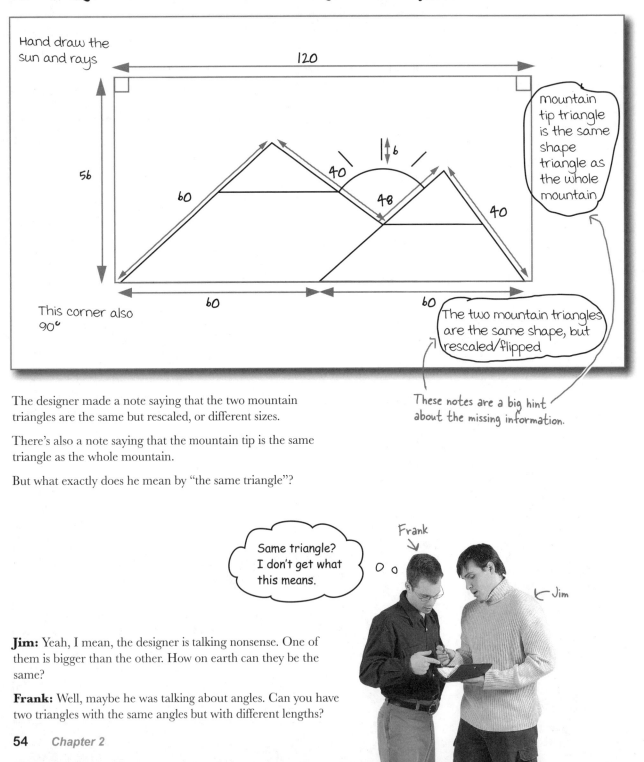

The designer made a note saying that the two mountain triangles are the same but rescaled, or different sizes.

There's also a note saying that the mountain tip is the same triangle as the whole mountain.

But what exactly does he mean by "the same triangle"?

Jim: Yeah, I mean, the designer is talking nonsense. One of them is bigger than the other. How on earth can they be the same?

Frank: Well, maybe he was talking about angles. Can you have two triangles with the same angles but with different lengths?

GEOMETRY CONSTRUCTION

Can you have two triangles with the same angles but with different lengths?

1) Cut or tear three narrow strips of paper, making them slightly different lengths.

2) Make them into a triangle and draw around it on some scrap paper.

3) Now fold each of your strips of paper in half, join them up to make a triangle again, and draw around it.

Compare your drawings to investigate what happens to the angles of a triangle when you make it bigger or smaller—you just need to make sure that you do the same to each side of your triangle.

Does your investigation help you to fill in any of the mystery angles on the design?

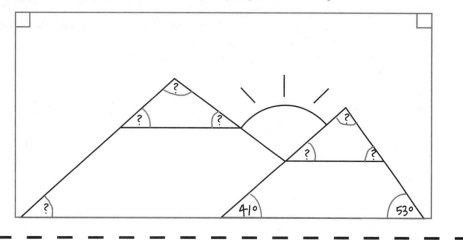

GEOMETRY CONSTRUCTION SOLUTION

Can you have two triangles with the same angles but with different lengths?

1) Cut or tear three narrow strips of paper, making them slightly different lengths.

2) Make them into a triangle and draw around it on some scrap paper.

3) Now fold each of your strips of paper in half, join them up to make a triangle again, and draw around it.

Compare your drawings to investigate what happens to the angles of a triangle when you make it bigger or smaller—you just need to make sure that you do the same to each side of your triangle.

Making a triangle bigger or smaller doesn't change the angles of the corners— providing you change the length of all the sides by the same ratio.

Does your investigation help you to fill in any of the mystery angles on the design?

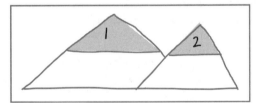

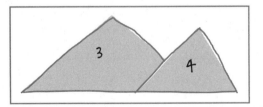

The mountains are basically made out of 4 of the same triangle in different sizes (one tucks in the back but the other corners are the same).

Changing the size of the triangle doesn't change the angles.... So each triangle must have the same angles as this one.

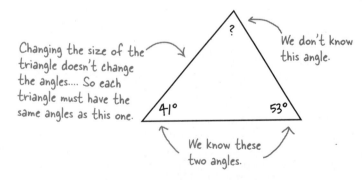

We don't know this angle.

?

41°   53°

We know these two angles.

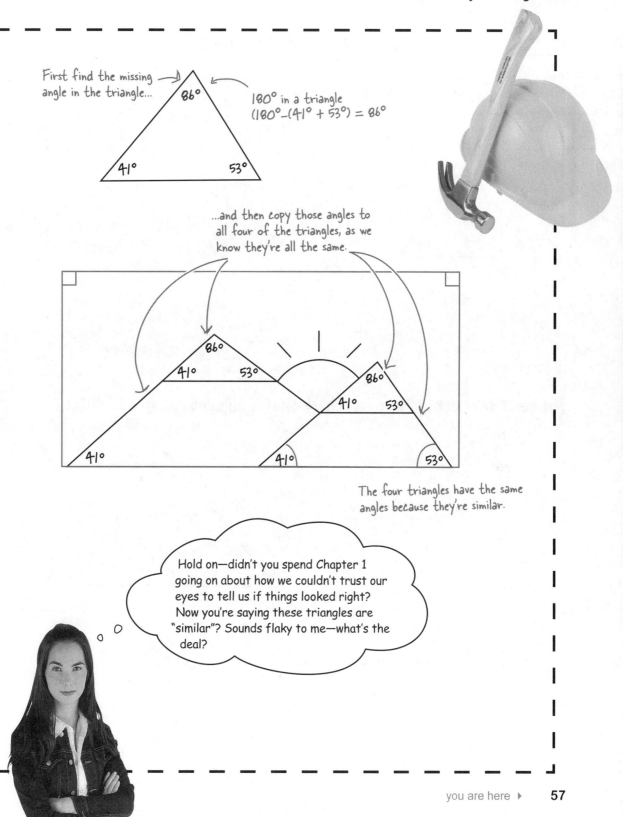

First find the missing angle in the triangle...

86°

41°   53°

180° in a triangle
(180°-(41° + 53°) = 86°

...and then copy those angles to all four of the triangles, as we know they're all the same.

86°
41°   53°

86°
41°   53°

41°

41°   53°

The four triangles have the same angles because they're similar.

Hold on—didn't you spend Chapter 1 going on about how we couldn't trust our eyes to tell us if things looked right? Now you're saying these triangles are "similar"? Sounds flaky to me—what's the deal?

o O

Similar triangles don't just look the same

Similarity is a key piece of geometry jargon. If two shapes are similar, then they don't just look alike, one is an exact scaled version of the other. This means that they have the same (equal) angles.

Similar triangles don't have to be the same way up.

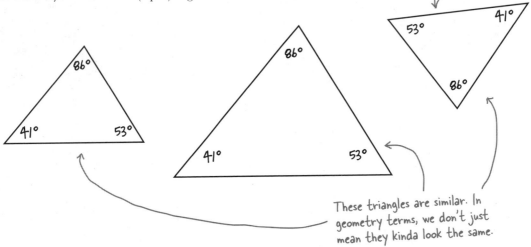

These triangles are similar. In geometry terms, we don't just mean they kinda look the same.

But don't take our word for it...check that it all adds up:

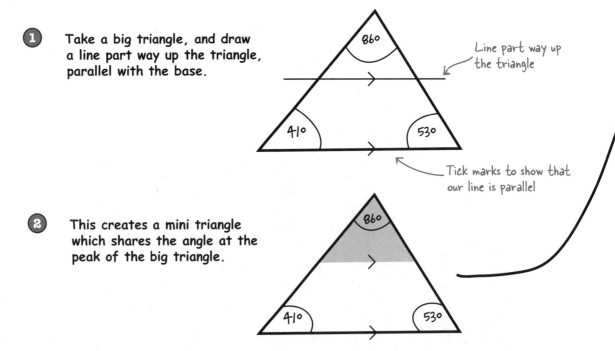

1 Take a big triangle, and draw a line part way up the triangle, parallel with the base.

Line part way up the triangle

Tick marks to show that our line is parallel

2 This creates a mini triangle which shares the angle at the peak of the big triangle.

Sharpen your pencil

Using what you know about parallel lines and the F pattern, find the two mystery angles to complete the mini triangle you've created.

Remember the F, N, Z patterns from chapter 1? No? Check out page 40.

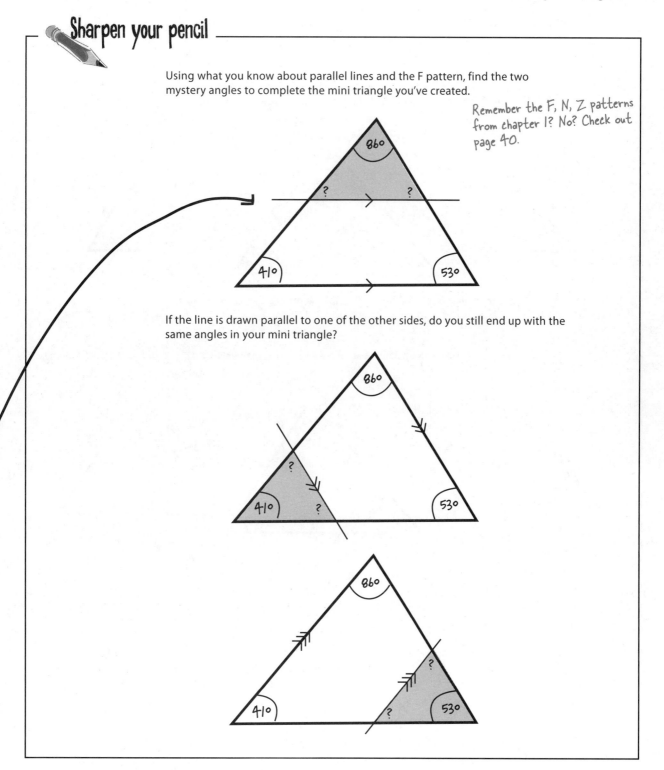

If the line is drawn parallel to one of the other sides, do you still end up with the same angles in your mini triangle?

Sharpen your pencil
Solution

Using what you know about parallel lines and the F pattern, find the two mystery angles to complete the mini triangle you've created.

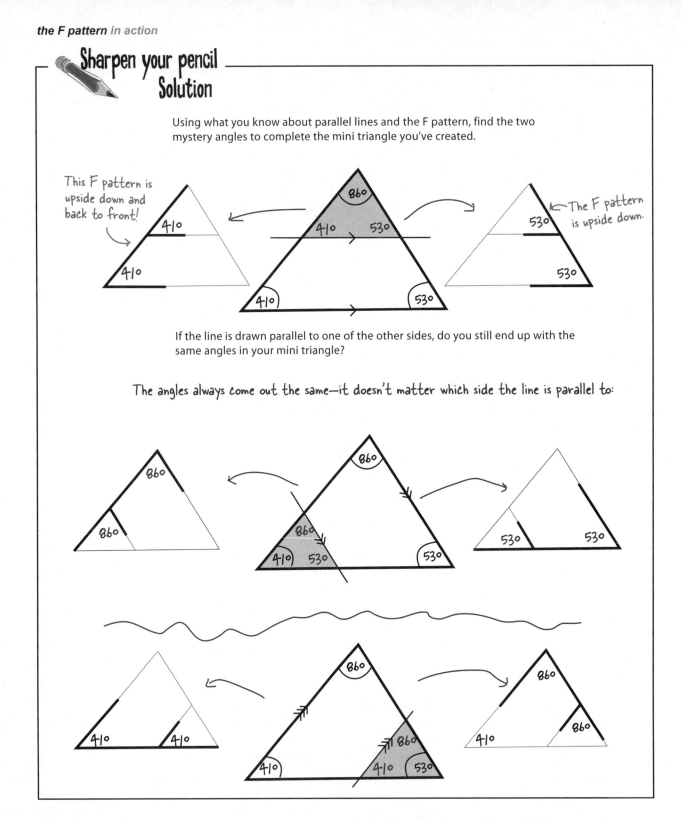

This F pattern is upside down and back to front!

The F pattern is upside down.

If the line is drawn parallel to one of the other sides, do you still end up with the same angles in your mini triangle?

The angles always come out the same—it doesn't matter which side the line is parallel to:

To use similarity, you need to be able to spot it

You won't normally get instructions that actually tell you that shapes are similar. So in order to use similarity, first you need to be able to be certain that shapes are similar.

You can do that by looking for matching sets of angles.

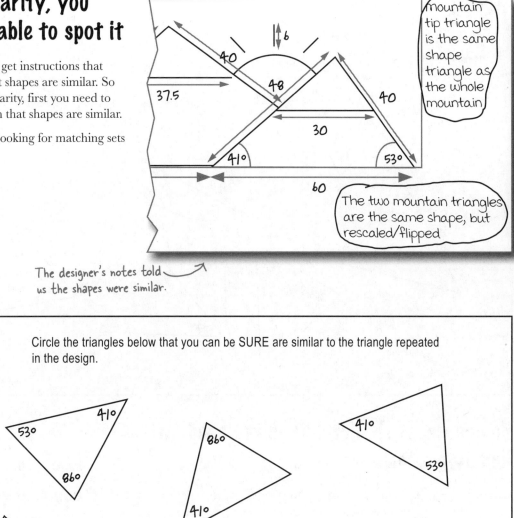

mountain tip triangle is the same shape triangle as the whole mountain

The two mountain triangles are the same shape, but rescaled/flipped

The designer's notes told us the shapes were similar.

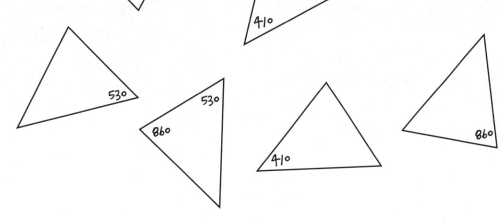

Exercise

Circle the triangles below that you can be SURE are similar to the triangle repeated in the design.

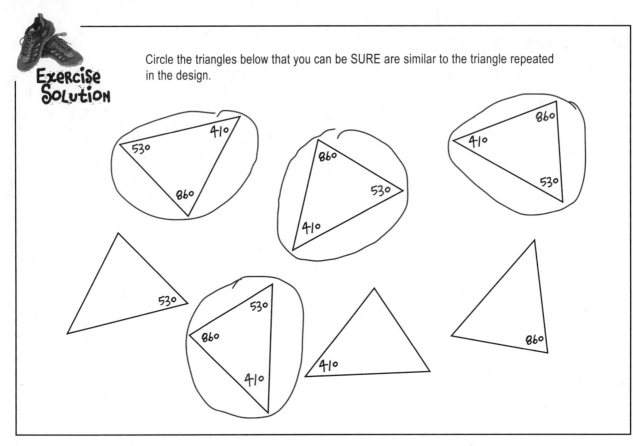

Exercise Solution

Circle the triangles below that you can be SURE are similar to the triangle repeated in the design.

You can spot similar triangles based on just <u>two</u> angles

We know that angles in a triangle add up to 180 degrees, so once you've got two angles in each triangle, you can always work out the third.

And if you've noticed that two angles in one triangle are equal to two angles in another triangle, then you can tell the triangles are similar without even doing any math!

Look for two equal angles to spot similar triangles.

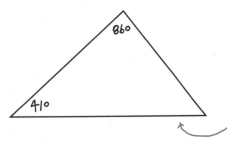

You can tell these are the same angle, even before you work out what that angle is.

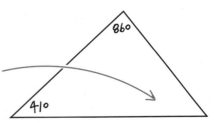

there are no
Dumb Questions

Q: What if the triangles are flipped, so one has a 41° on the right and the other has a 41° angle on left? Are they similar?

A: As long as you can spot another angle which is in both triangles then yes, they're definitely similar. Similarity is maintained even if your shape is reflected or rotated.

Q: Isn't using similarity kind of like cheating? Shouldn't I be working out all the angles individually?

A: Cheating? We like to think of it as working smarter rather than harder. It does save you plenty of leg work though. Most geometry teachers will be more impressed by use of similarity than repetitive calculations anyway—just make sure to make a note on your work that you used similarity.

Sharpen your pencil

How many repeated angles are there in total on this diagram, including the ones you've already marked plus the angles **a** through **i**? (Count each value once—if it's repeated don't count it again.)

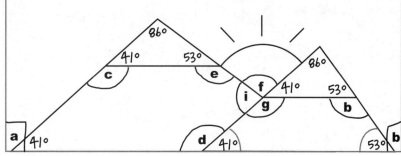

This corner also 90°

Sharpen your pencil
Solution

How many repeated angles are there in total on this diagram, including the ones you've already marked, plus the angles **a** through **i**? (Count each value only once—if it's repeated don't count it again.)

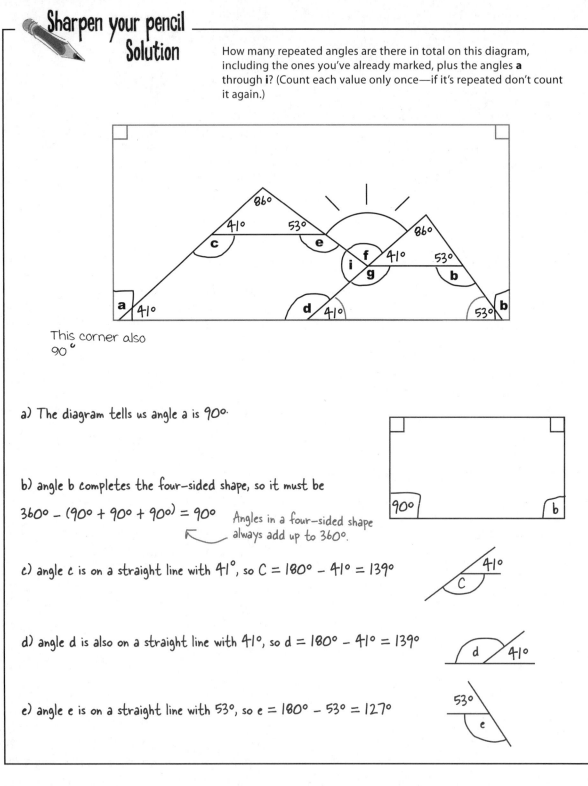

This corner also
90°

a) The diagram tells us angle a is 90°.

b) angle b completes the four-sided shape, so it must be

$360° - (90° + 90° + 90°) = 90°$ Angles in a four-sided shape always add up to 360°.

c) angle c is on a straight line with 41°, so C = 180° – 41° = 139°

d) angle d is also on a straight line with 41°, so d = 180° – 41° = 139°

e) angle e is on a straight line with 53°, so e = 180° – 53° = 127°

f) The left sides of the two mountains are both at 41°, so they must be parallel. This means that angle f makes a Z shape (alternate angles) with the 86° peak, so f must also be 86°.

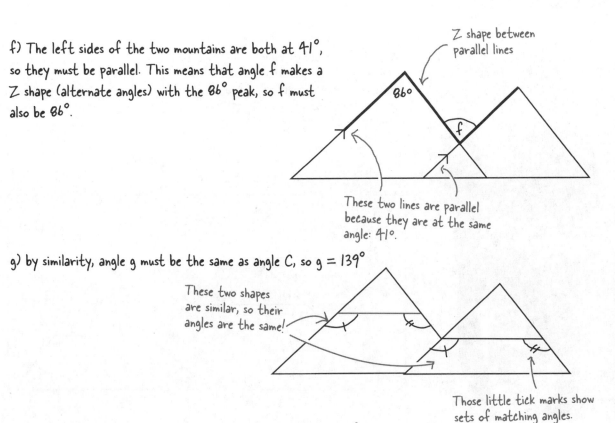

Z shape between parallel lines

86°

f

These two lines are parallel because they are at the same angle: 41°.

g) by similarity, angle g must be the same as angle C, so g = 139°

These two shapes are similar, so their angles are the same!

Those little tick marks show sets of matching angles.

h) by similarity, angle i must be the same as angle e, so i = 127°

i) angle i is on a straight line with f, which is 86°, so h = 180° − 86° = 94°

Different angles: 90°, 41°, 53°, 86°, 127°, 139°, 94°... = 7 different angles in total

Repeated angles: 90°, 41°, 53°, 86°, 127°, 139°... = 6 angles are repeated

A bunch are repeated, right? Were you surprised?

Employee of the month already?

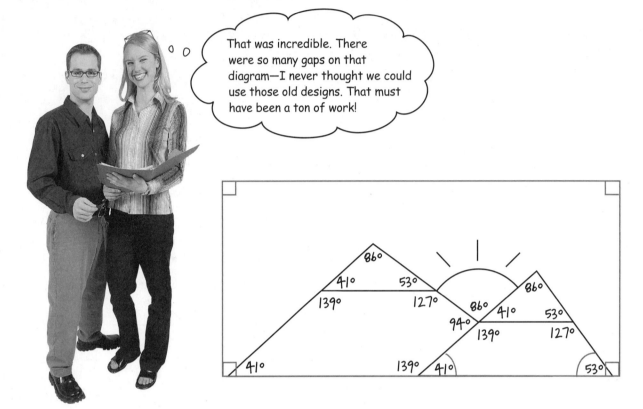

That was incredible. There were so many gaps on that diagram—I never thought we could use those old designs. That must have been a ton of work!

Similarity helps you work smarter, not harder, to find missing angles

While it's great to be able to use your Geometry Toolbox for finding missing angles, when you spot similarity, you can zip straight through to filling them in without even breaking out your calculation skills.

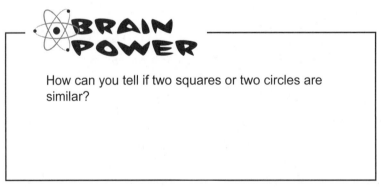

⚛ BRAIN POWER

How can you tell if two squares or two circles are similar?

Similarity Exposed

**This week's interview:
Similarity—faster or smarter—
which are you?**

Head First: Do you get a lot of pleasure out of saving people so much work? It must be nice to be liked.

Similarity: I do. It's why I do what I do, really. I'm a real stickler for recycling and conserving energy.

Head First: You mean like saving water?

Similarity: I mean like saving brain power! With similarity, you can do a calculation once and then reuse it over and over.

Head First: Oh, I see! Sorry…yes. So, what's the next step for you in your career?

Similarity: Well, really for me the next step is increasing recognition. I need people to know that I'm out here, waiting to get involved in saving them time and energy.

Head First: Well, that's certainly something we'd love to help you with. Thanks for the quick interview.

You sketch it—we'll etch it!

Now that you've figured out all of those missing angles, you've calculated everything that's needed to etch the design.

Fire up the etcher!

Now that we have all the angles we need, let's fire up the etcher and get the design permanently etched onto Liz's shiny new phone. How will it look?

But something's gone horribly wrong...

When you remove the phone from the etcher, the design isn't quite what you were expecting:

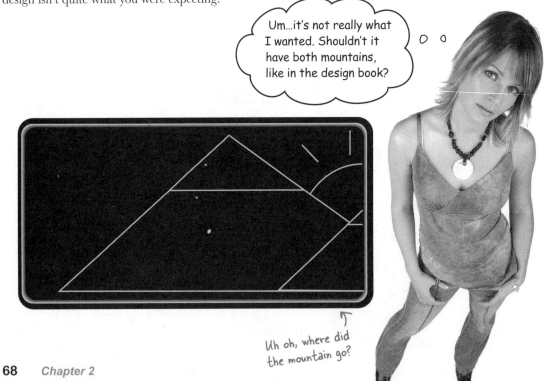

Um...it's not really what I wanted. Shouldn't it have both mountains, like in the design book?

Uh oh, where did the mountain go?

The boss isn't happy, but at least you're not fired...

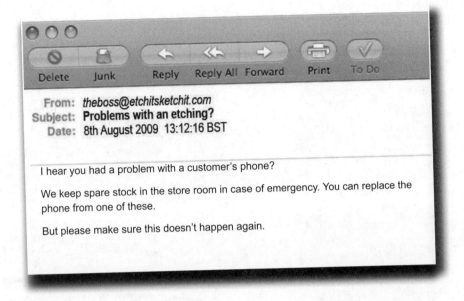

From: *theboss@etchitsketchit.com*
Subject: **Problems with an etching?**
Date: 8th August 2009 13:12:16 BST

I hear you had a problem with a customer's phone?

We keep spare stock in the store room in case of emergency. You can replace the phone from one of these.

But please make sure this doesn't happen again.

So what happened?

Here's what the design should have looked like. What do you think could have gone wrong?

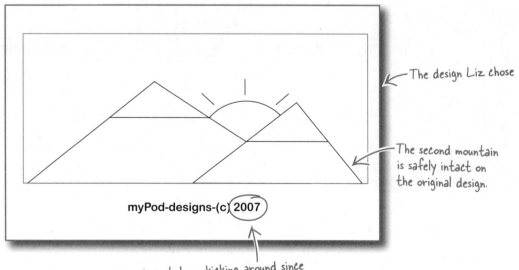

The design Liz chose

The second mountain is safely intact on the original design.

myPod-designs-(c) 2007

The design's been kicking around since 2007, so why is there a problem now?

It's a problem of scale...

The design Liz picked was originally created in 2007, when phones were a lot bigger than the one Liz has today. In fact, the design is *twice* as big as the space available on the back of Liz's phone for etching!

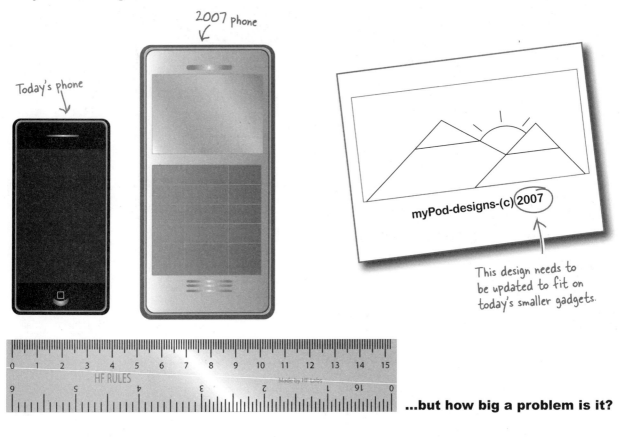

2007 phone

Today's phone

myPod-designs-(c) 2007

This design needs to be updated to fit on today's smaller gadgets.

...but how big a problem is it?

---

there are no Dumb Questions

Q: How come there weren't any units on the drawing? If they'd put units we could have seen whether it fit before we started etching.

A: You're right, we could have. In this case, though, the lengths on the diagram wouldn't be normal measures like millimeters or inches, but a special measure used by the etching machine.

Q: But shouldn't the drawing still have a scale? Wouldn't we be best to add one now?

A: For now we can just adjust the lengths we've got to work with—we can safely assume that whatever units the etcher uses, they aren't going to change before we etch again. But in the next chapter we're going to look at using scales in a lot more detail.

Pool Puzzle

Your job is to take steps from the pool that you think will help you make the design fit on Liz's new phone, and use them to complete your to-do list. (You don't need to use them all.)

1) Create a clean copy of the diagram with no numbers on it.
....

2)
....

3)
....

Note: each thing from the pool can only be used once!

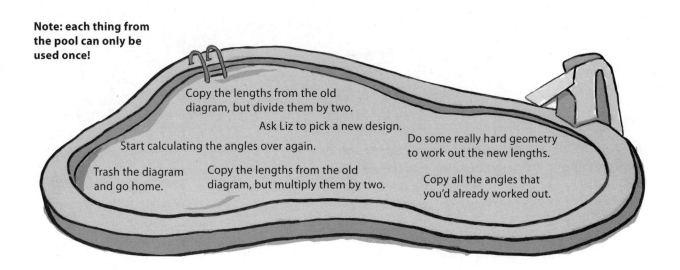

Copy the lengths from the old diagram, but divide them by two.

Ask Liz to pick a new design.

Start calculating the angles over again.

Do some really hard geometry to work out the new lengths.

Trash the diagram and go home.

Copy the lengths from the old diagram, but multiply them by two.

Copy all the angles that you'd already worked out.

Pool Puzzle Solution

Your job is to take from the pool the
steps you think will help you make
the design fit on Katie's new phone,
and use them to complete your
to-do list.

1) Create a clean copy of the diagram with no numbers on it.

2) Copy all the angles that you'd already worked out.

3) Copy the lengths from the old diagram, but divide them by two.

When you shrink
something evenly,
the angles don't
change.

And all the lengths
change by the same
FACTOR.

A factor is a common multiplier—
like if we doubled your pay and
also doubled your hours, we would
have increased both by a factor
of 2.

Ask Liz to pick a new design.

Start calculating the angles over again.

Do some really hard geometry
to work out the new lengths.

Trash the diagram
and go home.

Copy the lengths from the old
diagram, but multiply them by two.

Complex shapes can be similar, too

Similarity isn't just for triangles! Provided you shrink or grow your shapes **proportionally**, they can also be similar. When shapes are **proportional**, the **ratios** between the lengths of their different lines are the same.

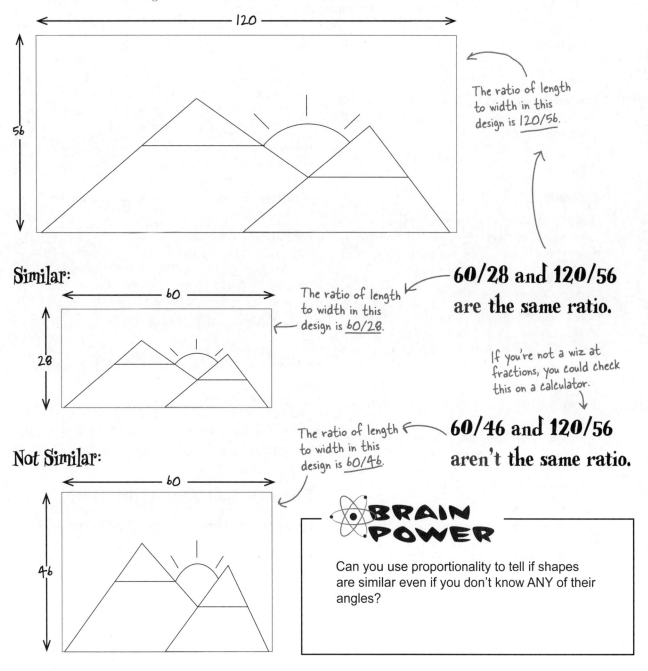

The ratio of length to width in this design is 120/56.

60/28 and 120/56 are the same ratio.

Similar:

The ratio of length to width in this design is 60/28.

If you're not a wiz at fractions, you could check this on a calculator.

60/46 and 120/56 aren't the same ratio.

Not Similar:

The ratio of length to width in this design is 60/46.

BRAIN POWER

Can you use proportionality to tell if shapes are similar even if you don't know ANY of their angles?

Similarity Exposed II

**This week's interview:
Ratios or angles, which is the
real similarity?**

Head First: You're really becoming popular—a lot of people are saying you're the time-saving technique they wish they'd always known.

Similarity: Yes—it's nice of you to say so! I do like to think I'm rather, um, efficient is the best word, I guess.

Head First: That's certainly true! But there's one thing I'm wondering….

Similarity: Go on….

Head First: Well, people recognize you by matching angles—and others use the proportional thing—and I'm just wondering, which is the *real* you?

Similarity: I don't understand. You mean you think I can only be one or the other?

Head First: Well, surely one is what you're really about, and the other is just a convenient alternative way of presenting yourself. I want to get to the heart of the real similarity—who are you when you're just relaxing at home?

Similarity: Well, to be honest, I really am always both! I know it sounds silly, but I've never thought of my different aspects as being separate. With triangles, and a lot of other shapes, too, if the angles are matching, then the sides are also proportional. I can't really pick and choose one or the other!

Head First: And what about if you've got proportionality; if ratios between the lengths of a triangle are the same, but you don't have matching angles? Do you feel something is missing?

Similarity: But that could never happen with a triangle! That's just how it is. Anytime triangles have the same ratios, they have the same angles. You've made me anxious now…but honestly, it's just not possible. Proportionality and angles—with triangles it's always about both, equally together!

Head First: Together? I didn't know you were mixing it up like that. Interesting…. Now, you said, "a lot of other shapes, too"—that suggests that it's not always the case that angles and proportionality go together?

Similarity: Ah, well, there are some shapes that are different. Take rectangles for example. All rectangles have the same angles—90, 90, 90, and 90 degrees. But they aren't all proportional—you can have long skinny ones and short fat ones.

Head First: So you don't work with rectangles at all?

Similarity: Oh, I do. But only proportional ones. Like if you had a rectangle with sides 3 and 6, and one with 4 and 8—you'd know they were similar. And squares! I love squares. All of them are similar. Every single one. Beautiful. Just beautiful.

Head First: Right. Beautiful squares, eh? Thanks for the interview.

You can spot similarity using angles or the ratios between lengths or sides, or both.

Sharpen your pencil

Based on the old diagram and the angles you'd figured out earlier, mark up a fresh design to fit Liz's phone. It needs to be half the size of the original.

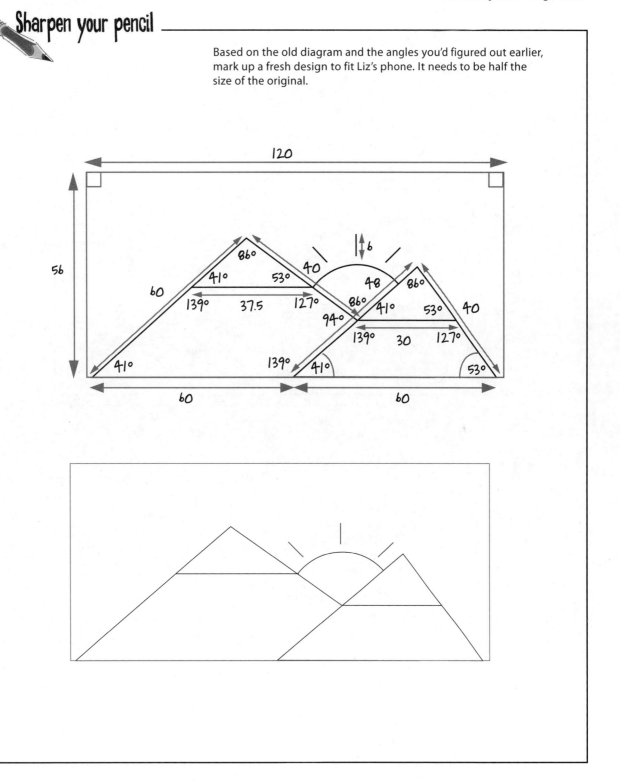

Sharpen your pencil
Solution

Based on the old diagram and your angle workings, mark up a fresh design to fit Liz's phone. It needs to be half the size of the original.

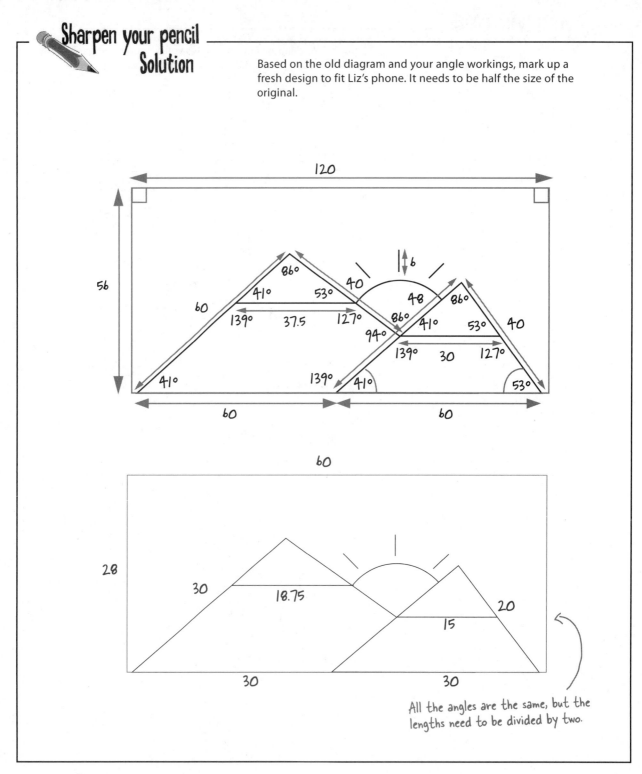

All the angles are the same, but the lengths need to be divided by two.

You sketch it—we'll etch it (to fit)

Liz is a very patient customer, she's hung around while you used similarity and proportionality to resize the design to fit perfectly on her new replacement phone. It does look good though!

I love it, thanks!

BULLET POINTS

- **Similar** shapes have the **same angles.**

- Similar shapes have the **same ratios** between **lengths of sides.**

- Similar triangles have the same three angles (and you can tell from **just two**).

- Some shapes are always similar.

- All **circles** are similar.

- All **squares** are similar.

Just try drawing a square which doesn't have four angles all 90° and all sides the same ratio: 1/1.

Liz is back with a special request

The great thing about happy customers is that they just keep coming back. Impressed by the effort you put in to getting her phone just right, Liz is trusting you with another great gig.

Hey! Everyone thinks my phone looks great. Can you do a totally custom design on my brother's iPod of his band's logo?

All the arrows are the same.

The square part is half the length of the arrow head part.

Triangle sides are the same.

The darker set of arrows are 3/4 size.

The sketch is pretty...er...sketchy. Drawn on the back of a flyer by the drummer.

An "arrow" is one of these.

Sharpen your pencil

Before you start sketching the design, what lengths and angles do you need to find? Could you use similarity to save yourself some time and effort?

Sharpen your pencil
Solution

Before you start sketching the design, which lengths and angles do you need to find? Could you use similarity to save yourself some time and effort?

We need to find the lengths and angles of the sides of one of the small triangles and one of the big triangles, plus the lengths of the sides of one of the small squares and one of the big squares.

Although there aren't any length or angle markings on the diagram, the instructions give us plenty of clues—and there's a ton of similarity going on here.

The diagram is made up of six similar arrow shapes like this.

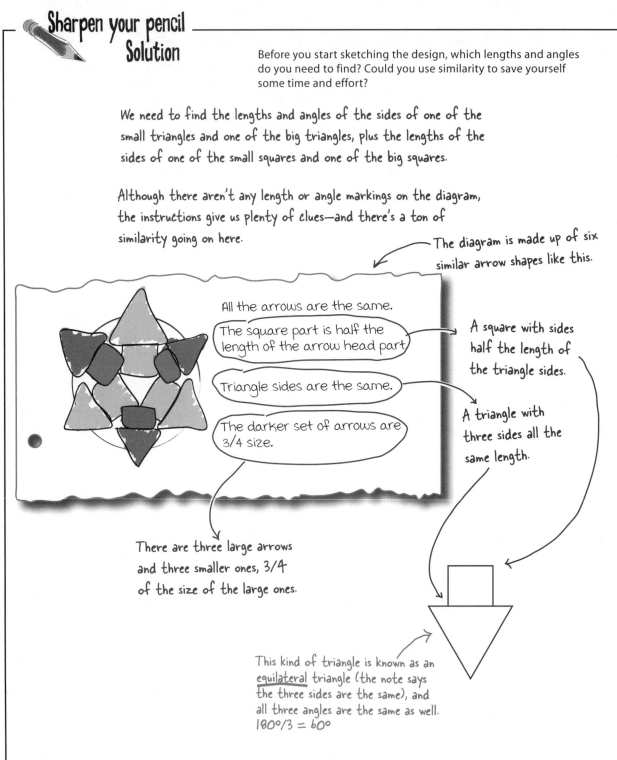

All the arrows are the same.

The square part is half the length of the arrow head part

Triangle sides are the same.

The darker set of arrows are 3/4 size.

A square with sides half the length of the triangle sides.

A triangle with three sides all the same length.

There are three large arrows and three smaller ones, 3/4 of the size of the large ones.

This kind of triangle is known as an equilateral triangle (the note says the three sides are the same), and all three angles are the same as well. 180°/3 = 60°

Similar shapes that are the same size are congruent

Shapes that are similar have equal angles and are proportional, but if they're actually the exact same size, then we say that they're **congruent**.

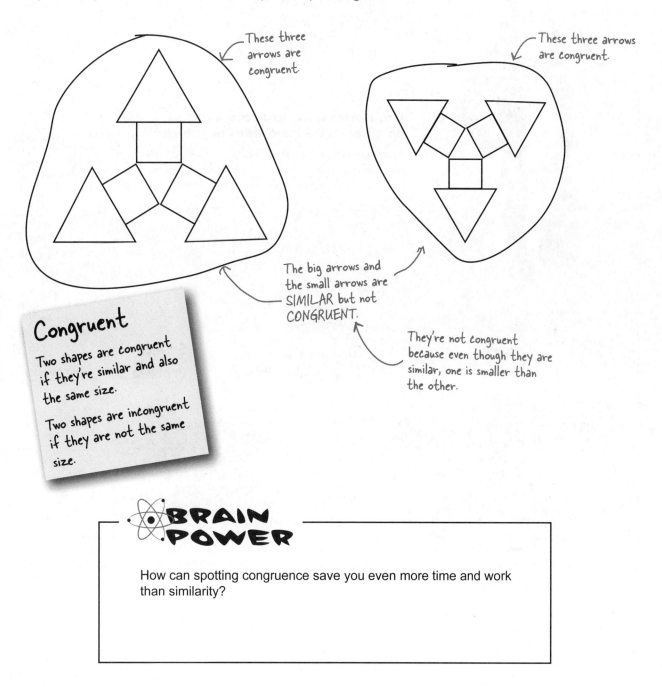

These three arrows are congruent.

These three arrows are congruent.

The big arrows and the small arrows are SIMILAR but not CONGRUENT.

They're not congruent because even though they are similar, one is smaller than the other.

Congruent

Two shapes are congruent if they're similar and also the same size.

Two shapes are incongruent if they are not the same size.

⚛ BRAIN POWER

How can spotting congruence save you even more time and work than similarity?

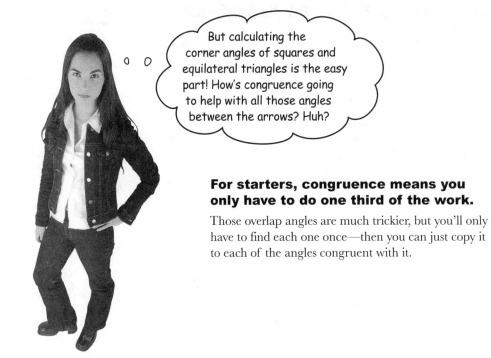

For starters, congruence means you only have to do one third of the work.

Those overlap angles are much trickier, but you'll only have to find each one once—then you can just copy it to each of the angles congruent with it.

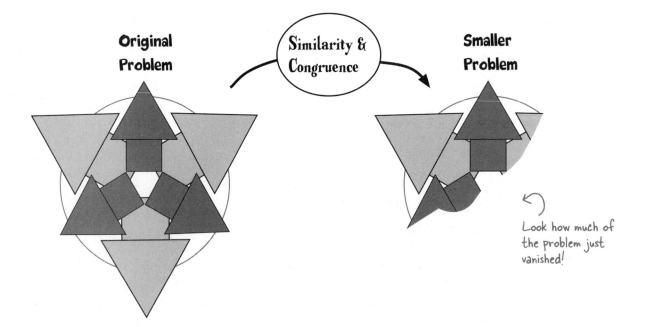

Use what you know to find what you don't know

We said it in Chapter 1, and it still applies now. Work from what you do know to find out what you don't know. Like the angles **between** the arrows.

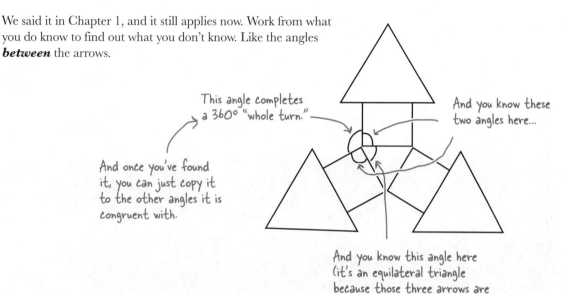

This angle completes a 360° "whole turn."

And once you've found it, you can just copy it to the other angles it is congruent with.

And you know these two angles here...

And you know this angle here (it's an equilateral triangle because those three arrows are congruent).

And if you don't have what you need, add it!

You can add parallel or perpendicular lines to your sketch to break down the missing angles into parts you have the tools to find.

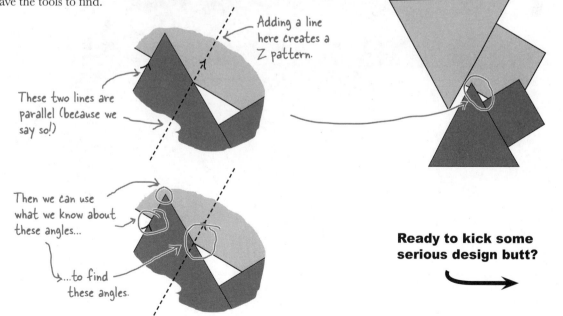

Adding a line here creates a Z pattern.

These two lines are parallel (because we say so!)

Then we can use what we know about these angles...

...to find these angles.

Ready to kick some serious design butt?

Exercise

There are 60 angles on the band's logo design. Use the space on the right to start working on the sketch and calculate them all. How many of each different angle are there?

Feeling overwhelmed?
Don't panic! Everything
you need is in your
Chapter 1 toolbox.

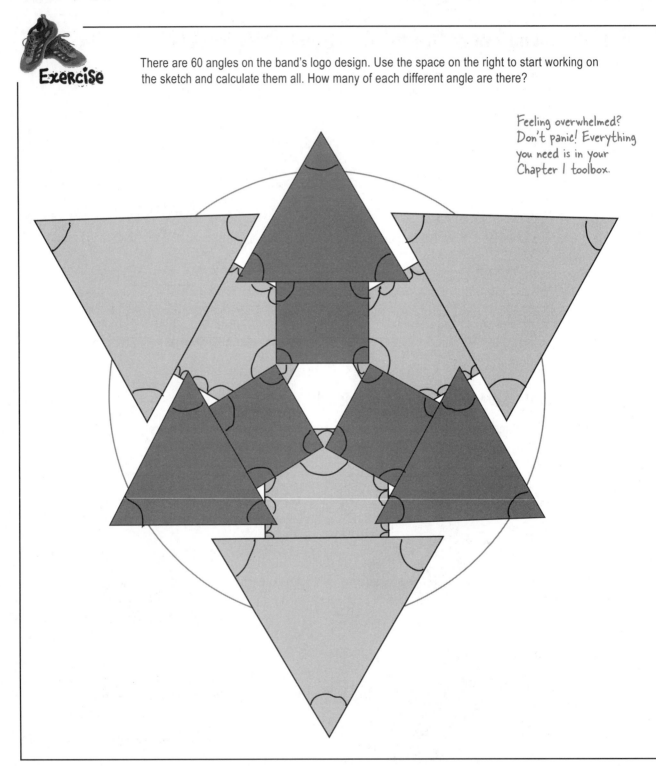

Exercise Solution

There are 60 angles on the band's logo design. Calculate them all. How many of each different angle are there?

There are 21 angles that are 60°, 24 angles that are 90°, 9 angles that are 120°, and 6 that are 150°.

Here's how you can find them all:

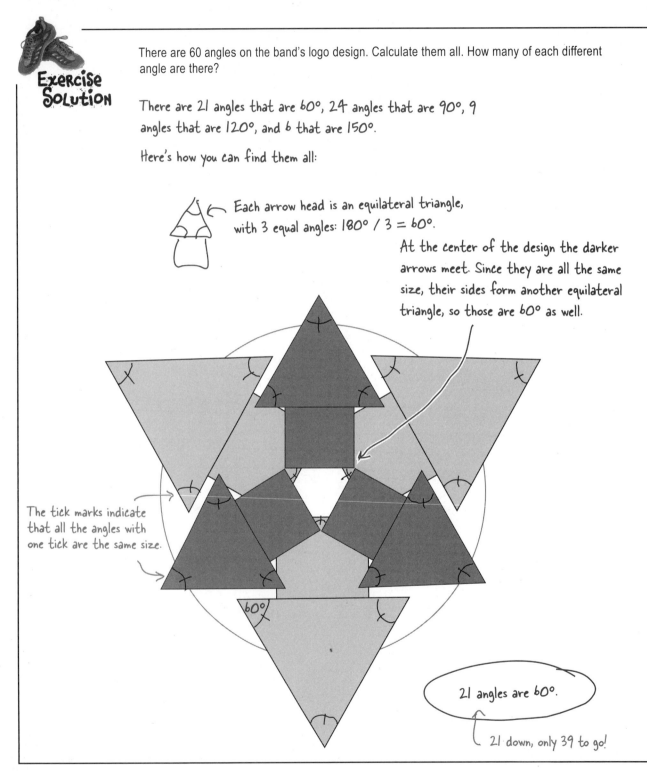

Each arrow head is an equilateral triangle, with 3 equal angles: 180° / 3 = 60°.

At the center of the design the darker arrows meet. Since they are all the same size, their sides form another equilateral triangle, so those are 60° as well.

The tick marks indicate that all the angles with one tick are the same size.

60°

21 angles are 60°.

21 down, only 39 to go!

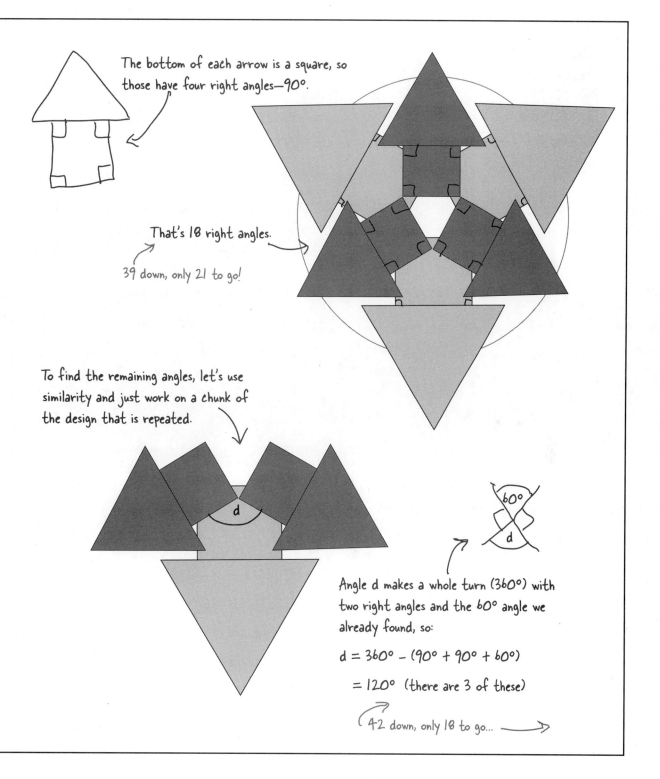

The bottom of each arrow is a square, so those have four right angles—90°.

That's 18 right angles.

39 down, only 21 to go!

To find the remaining angles, let's use similarity and just work on a chunk of the design that is repeated.

Angle d makes a whole turn (360°) with two right angles and the 60° angle we already found, so:

d = 360° – (90° + 90° + 60°)

= 120° (there are 3 of these)

42 down, only 18 to go...

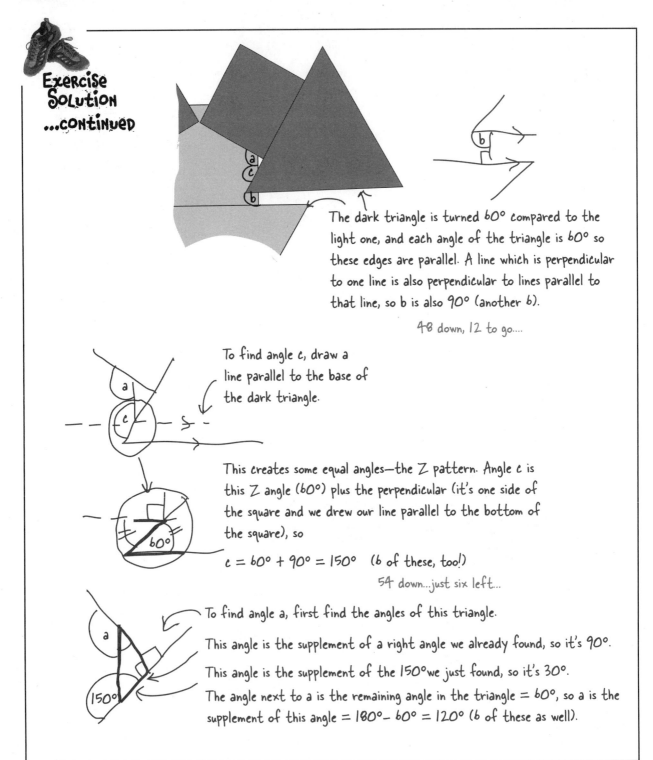

Exercise
Solution
...continued

The dark triangle is turned 60° compared to the light one, and each angle of the triangle is 60° so these edges are parallel. A line which is perpendicular to one line is also perpendicular to lines parallel to that line, so b is also 90° (another b).

48 down, 12 to go....

To find angle c, draw a line parallel to the base of the dark triangle.

This creates some equal angles—the Z pattern. Angle c is this Z angle (60°) plus the perpendicular (it's one side of the square and we drew our line parallel to the bottom of the square), so

c = 60° + 90° = 150°   (6 of these, too!)

54 down...just six left...

To find angle a, first find the angles of this triangle.

This angle is the supplement of a right angle we already found, so it's 90°.

This angle is the supplement of the 150°we just found, so it's 30°.

The angle next to a is the remaining angle in the triangle = 60°, so a is the supplement of this angle = 180°– 60° = 120° (6 of these as well).

Fireside Chats

Tonight's talk: **Similarity and Congruence**

Similarity:

I'm worried about you. You seem neurotic, always concerned about being a size zero, or seven, or whatever. You need to relax more.

Congruence:

But size DOES matter. It's all very well having your angles right on paper, but if something is too big or too small, in the real world size actually matters.

I just think if you stopped worrying so much about size and focused on proportionality you'd have more opportunities. I get used all the time—my flexibility is a real asset.

And potentially a real headache! Sometimes it doesn't cut it just being similar—and that's where I come in, when you need to rely on size as well as shape. Plus, I'm faster to use.

Faster? How? I'm pretty fast, you know!

Yeah, you're quick, but even if something's similar, if it's not congruent, then you've gotta do some math for the lengths. OK, it's only multiplication and division, but it all takes time.

True. And I'm a big stickler for efficiency. People would do well to use both of us more!

Yes. Are you getting much non-triangle work these days? I'm mostly getting triangle stuff still, and it's not that I don't like it…but you know, I have so much more potential.

Oh, tell me about it! I think they forget that we're not all about the triangles. And that we travel so well.…

Completely—I love to travel! You can flip me upside down, back to front, spin me around and move me from one place to another, and I still work just as well.

Great to catch up! I'll let you know if I get any work you could help out with.

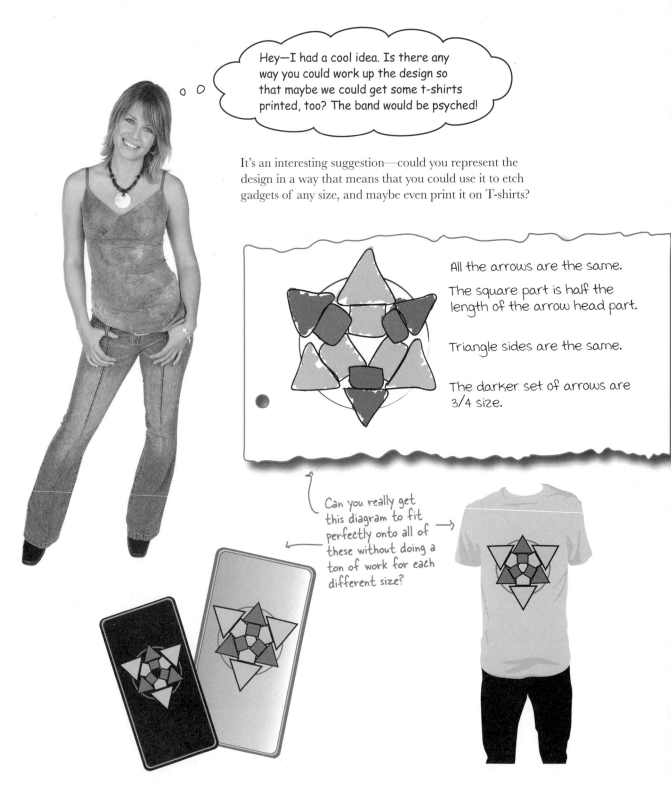

Hey—I had a cool idea. Is there any way you could work up the design so that maybe we could get some t-shirts printed, too? The band would be psyched!

It's an interesting suggestion—could you represent the design in a way that means that you could use it to etch gadgets of any size, and maybe even print it on T-shirts?

All the arrows are the same.

The square part is half the length of the arrow head part.

Triangle sides are the same.

The darker set of arrows are 3/4 size.

Can you really get this diagram to fit perfectly onto all of these without doing a ton of work for each different size?

I think you're getting a bit ahead of yourself here—we don't have any lengths at all yet. It's way too early to be thinking about resizing it...

BRAIN BARBELL

It would be cool to finish marking up your design of the band's logo so that it would fit on any size gadget—even a T-shirt.

But what does that actually mean you need to do?

a) Draw the design at different sizes and mark up different lengths on each of them, that way you'll always have one to fit.

Write what you'd need to do and why you do or don't think its the best approach.

....

....

....

b) Forget the final size, and just make sure you've captured the relationships between the lengths of the lines, so it stays proportional.

....

....

....

BRAIN
BARBELL SOLUTION

It would be cool to finish marking up your design of the band's logo so that it would fit on any size gadget—even a T-shirt.

But what does that actually mean you need to do?

a) Draw the design at different sizes and mark up different lengths on each of them, that way you'll always have on to fit.

Apart from taking forrrevvver, this still wouldn't guarantee that you always had a ✗

size that fit...what if they started making an XXL t-shirt? You'd have to fiddle

with the numbers again.

b) Forget the final size, and jut make sure you've captured the ratios between the lengths of the lines, so it stays proportional.

By using ratios rather than actual sizes, you can make your design more flexible. You'll still ✓

need to do a little bit of math, but nothing major.

there are no
Dumb Questions

Q: Surely even if I use the proper sizes, as long as it's proportional, I can just do some more math to work out a different size by multiplying by a new factor?

A: You can—it's just that the math will be a lot harder and the ratios much less clear. Technically it's not wrong, it's just not the simplest way to go about creating a scalable design.

Ratios can be more useful than sizes

A ratio captures the proportions of a shape, and then by using a different factor, we can create a similar shape of any size.

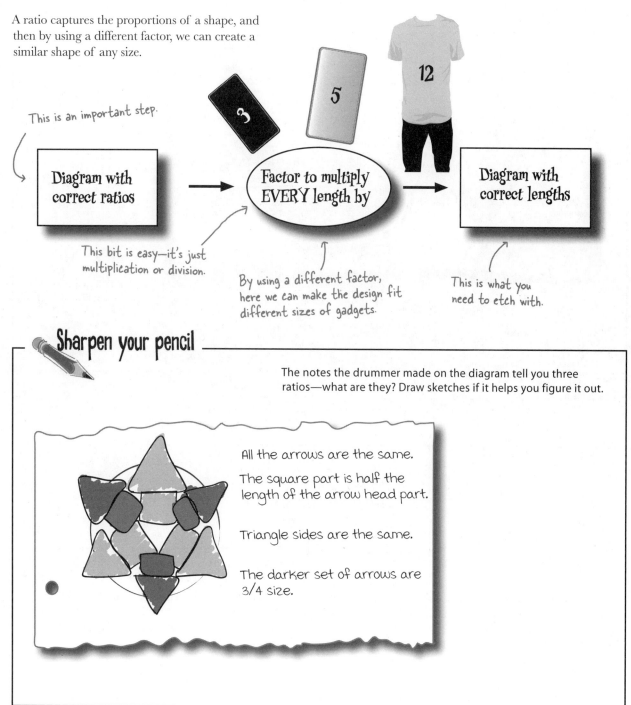

This is an important step.

| Diagram with correct ratios |

This bit is easy—it's just multiplication or division.

3

5

12

| Factor to multiply EVERY length by |

By using a different factor, here we can make the design fit different sizes of gadgets.

| Diagram with correct lengths |

This is what you need to etch with.

Sharpen your pencil

The notes the drummer made on the diagram tell you three ratios—what are they? Draw sketches if it helps you figure it out.

All the arrows are the same.

The square part is half the length of the arrow head part.

Triangle sides are the same.

The darker set of arrows are 3/4 size.

Sharpen your pencil Solution

The notes the drummer made on the diagram tell you three ratios—what are they? Draw sketches if it helps you figure it out.

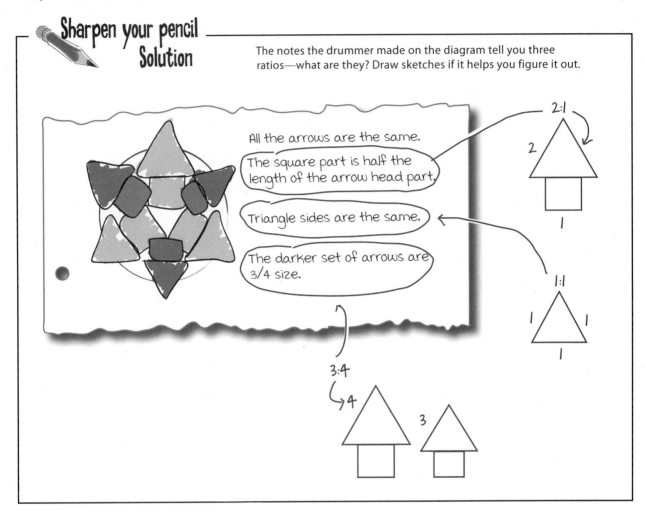

All the arrows are the same.

The square part is half the length of the arrow head part.

Triangle sides are the same.

The darker set of arrows are 3/4 size.

2:1

1:1

3:4

---

there are no Dumb Questions

Q: How come 1:1 is a ratio? Isn't that a bit pointless?

A: It does sound a bit weird, doesn't it? 1:1 is a way of indicating that it's exactly the same size as the original. If you didn't put the 1:1 in there, then someone reading your work or diagram might wonder whether you'd forgotten to put a ratio in for that item. 1:1 indicates clearly that it's the same size.

Q: The drummer wrote that the dark arrows are 3/4 size, so why have you written it as 3:4?

A: 3/4 and 3:4 are just different ways of indicating the same relative proportion. 3/4 is fraction notation and 3:4 is ratio notation. Unless you've been specifically told to use one or the other then they're mostly swappable.

Uh, sorry, dude, but you've messed up. You're saying that the arrow heads are triangles with size 3, and size 4, and size 2...well, they can't be ALL of them!

Frank

Jim

Joe

Frank: I kinda followed what you did though…with the ratios, and it all makes sense.

Jim: I don't think so, it has to be wrong. There's no way it can be 2 and 4, or 2 and 3…not at the same time. You need to take another look.

Joe: That or the drummer got it wrong. They're not always the brightest in the band.…

Frank: Could it be something to do with it being ratios rather than sizes? Like, I'm twice as old as my sister, but I'm also half as old as my dad.…

Joe: And you're half as good-looking as me. What are you going on about? Those are both twos…we're worried about different numbers!

Frank: Yeah, but my dad is four times as old as my kid sister—so, you could say their ages were 4:1, even though my dad and I are 2:1. They're both true; it's just relative.

Jim: Relative to what, though?

Frank: That's the thing—maybe we just need to choose one thing for our ratios to be relative to and then stick with it.

Ratios need to be consistent

The ratios we took from the diagram are individually correct, but they describe different relationships between the lengths of our lines.

On our design, we need to make sure that we reflect all the ratios at the same time, which means we have to pick one thing and then work everything out relative to that.

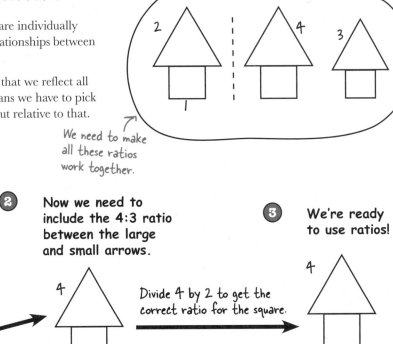

We need to make all these ratios work together.

1 Let's choose this biggest arrow to be the thing we're working relative to.

2 Now we need to include the 4:3 ratio between the large and small arrows.

Divide 4 by 2 to get the correct ratio for the square.

Divide 3 by 2 to get the correct ratio for the square.

3 We're ready to use ratios!

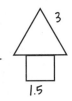

That decimal looks like it could be tricky; if we picked some different numbers, could we get rid of it?

While the decimal isn't really a problem, it's certainly easier to work with ratios that are only whole numbers: your brain can compare ratios like 3:4 and 7:8 in a way that you probably can't just figure out which is bigger out of 2.67:5.3 and 4.56:6.2.

Sharpen your pencil

The ratios have turned out to be 4, 3, 2, and 1.5.

What would be the smallest set of whole numbers you could substitute and still keep the ratios the same?

Use your answers from here in this answer.

Exercise

Time to get etching! If the design will fit on Liz's brother's iPod with the biggest arrow head edges at 2.4cm long, what lengths do the other lines—a, b, c—need to be?

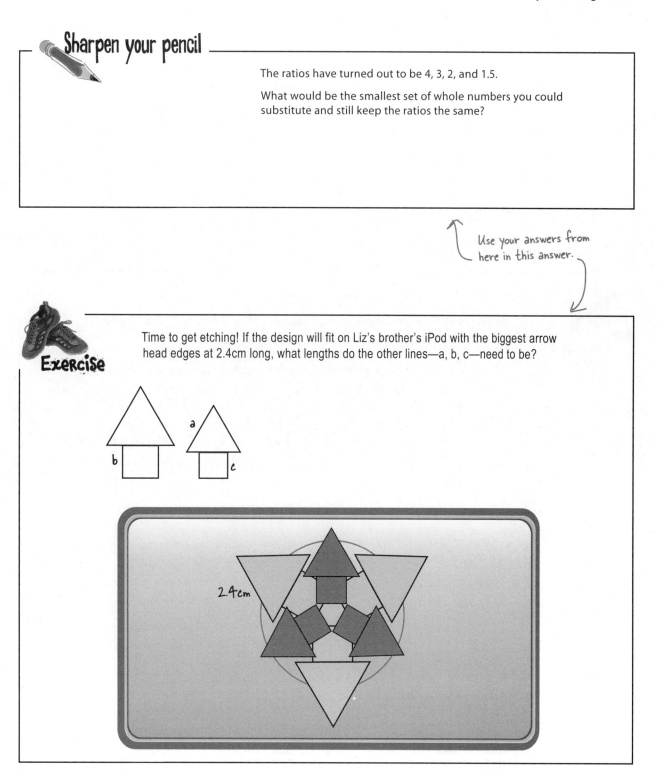

double *it up*

Sharpen your pencil Solution

The ratios have turned out to be 4, 3, 2, and 1.5.

What would be the smallest set of whole numbers you could substitute and still keep the ratios the same?

If we multiply all the ratios by 2, then we get a set of whole numbers with the same ratios: 8, 6, 4, and 3.

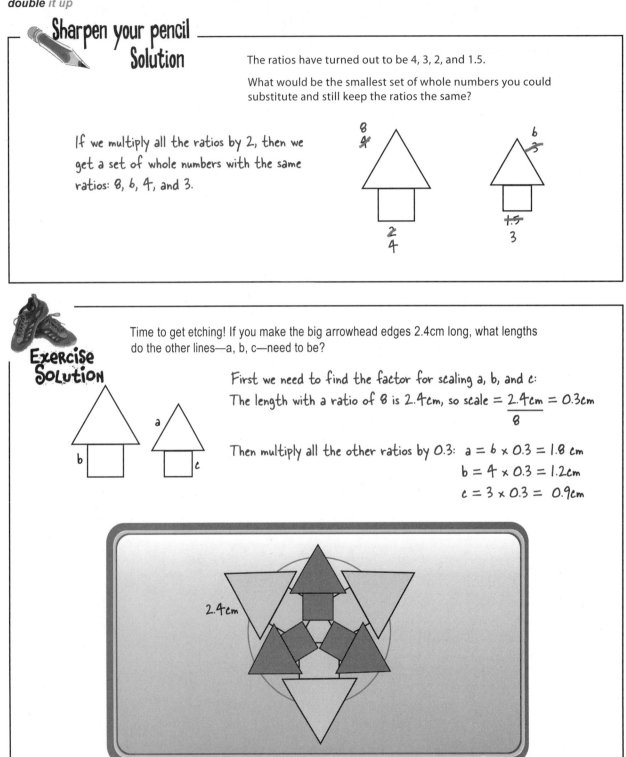

Exercise Solution

Time to get etching! If you make the big arrowhead edges 2.4cm long, what lengths do the other lines—a, b, c—need to be?

First we need to find the factor for scaling a, b, and c:

The length with a ratio of 8 is 2.4cm, so scale $= \dfrac{2.4cm}{8} = 0.3cm$

Then multiply all the other ratios by 0.3:

$a = 6 \times 0.3 = 1.8 \text{ cm}$

$b = 4 \times 0.3 = 1.2 \text{cm}$

$c = 3 \times 0.3 = 0.9 \text{cm}$

Your new design ROCKS!

Check out the band's latest blog entry on your brand
new iPhone—your myPod bonus for getting that
extra order to print all those T-shirts!

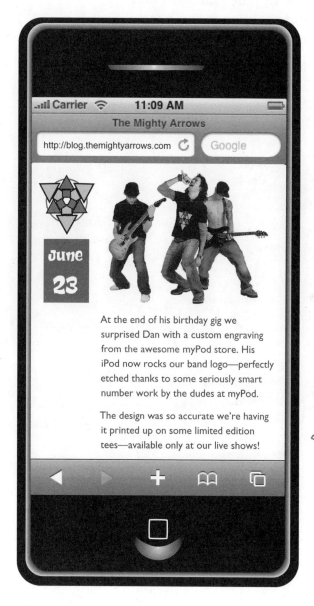

Oh, yeah, you got backstage tickets, too!

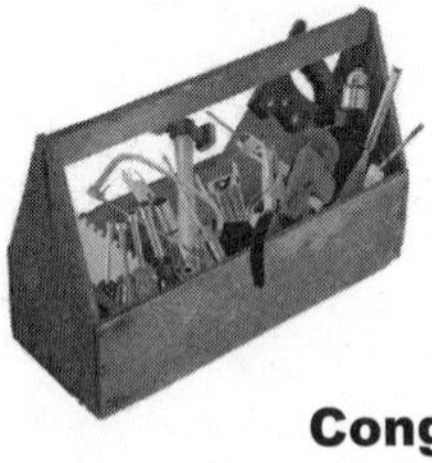

Your Geometry Toolbox

**You've got Chapter 2
under your belt and now
you've added Similarity and
Congruence to your toolbox. For a
complete list of tool tips in the book,
visit www.headfirstlabs.com/geometry.**

Similar shapes have
equal angles.

Similar shapes have
the same ratios
between side lengths.

You can spot similar triangles
from two equal angles.

Actually they have three
angles that are equal, but
as soon as you spot two you
know that the third is gonna
match, too.

Congruent shapes are
similar and the same size.

Congruent shapes have
angles that are equal, just
like similar shapes, but
they're also the same scale.

All squares and
circles are similar.

There's some major construction projects up ahead. And there's been a collision between letters and numbers!

Letters and numbers colliding? Oh, no! It must be ALGEBRA!

But DON'T TURN BACK!

Instead, check out our FREE (and pain free) downloadable bonus chapter on algebra and geometry here:

http://www.headfirstlabs.com/geometry

...and find out how YOU can harness the power of algebra to solve 215 geometry problems in one go.

3 the pythagorean theorem

All the right angles

You just straighten one leg, then straighten the other leg, clap your hands, and you're dancing the hypotenuse! It's all in the hips, really.

Sometimes, you really need to get things straight.

Ever tried to eat at a wobbly table? Annoying, isn't it? There is an alternative to shoving screwed-up paper under the table leg though: use the Pythagorean Theorem to make sure your designs are **dead straight** and not just *quite straight*. Once you know this pattern, you'll be able to **spot and create right angles** that are **perfect every time.** Nobody likes to pick spagetti out of their lap, and with the **Pythagorean Theorem**, you don't have to.

Giant construction-kit skate ramps

Sam is seriously into skating, and she funds her skate habit by building street courses—a jungle of ramps and rails where skaters can kickflip and 5-0 themselves silly.

But building the ramps takes up time (and money) that Sam would rather spend on actually skating. But last week, while babysitting her nephew, Sam had the best idea she's ever had: **use quick assembly standard sized pieces to build the skate ramps.**

..and skateboards, and shoes and hoodies....

With giant construction kit parts, instead of making all our ramps from scratch, we could use standard size pieces that assemble together quickly. It'll be faster and cheaper!

The inspiration for Sam's genius plan was a kids' construction kit.

Sam

Standard-sized-quick-assembly-what?!?

Sam's found a supplier for giant construction kit parts. Kwik-klik makes parts which are just like construction kit toys you might have seen or played with when you were younger, but on a much bigger scale.

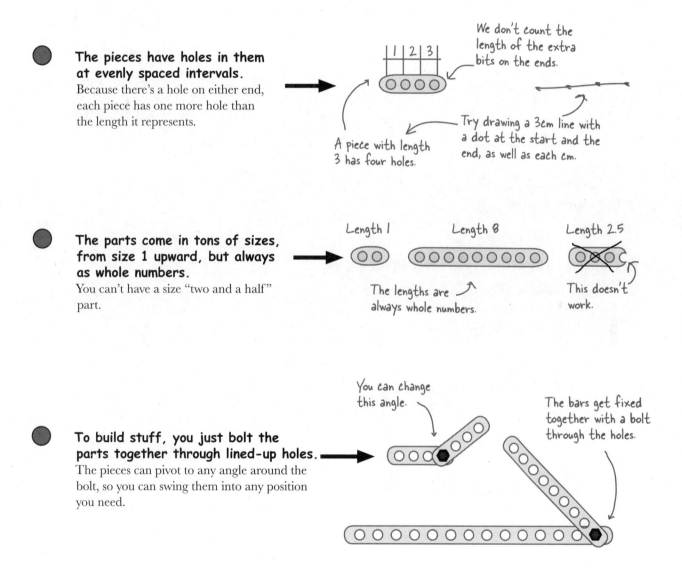

The pieces have holes in them at evenly spaced intervals.
Because there's a hole on either end, each piece has one more hole than the length it represents.

We don't count the length of the extra bits on the ends.

A piece with length 3 has four holes.

Try drawing a 3cm line with a dot at the start and the end, as well as each cm.

The parts come in tons of sizes, from size 1 upward, but always as whole numbers.
You can't have a size "two and a half" part.

Length 1   Length 8   Length 2.5

The lengths are always whole numbers.

This doesn't work.

To build stuff, you just bolt the parts together through lined-up holes.
The pieces can pivot to any angle around the bolt, so you can swing them into any position you need.

You can change this angle.

The bars get fixed together with a bolt through the holes.

So easy, anyone can do it! But Sam's first ramp has a major problem...

The ramps must have <u>perpendicular</u> uprights ← The "upright" is the vertical part of the ramp.

Sam needs the ramps to have true vertical uprights, **perpendicular** (90°) to the bases, so that she can put them back to back or against a wall without a nasty gap to get your wheels caught in.

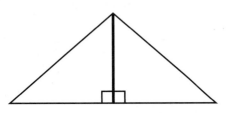 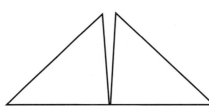

Perpendicular uprights are perfectly vertical and let the ramps fit together.

Slanted uprights create wheel–trapping gaps.

Lines or objects that are perpendicular meet or cross at a right angle.

But Sam's first prototype is not squaring up

Even though it looked good on paper, now Sam's built her first ramp and…well, it's just plain wonky.

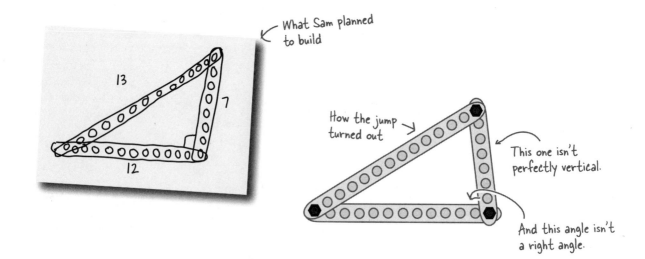

← What Sam planned to build

13

7

12

How the jump turned out ⟩

This one isn't perfectly vertical.

And this angle isn't a right angle.

Wow, I guess it's important which lengths I put together! It would be cool to know that the ramp will work out straight BEFORE we buy parts and build it....

BRAIN POWER

Could you use a pencil, a ruler and some paper to check out whether a ramp design will give you a perfectly vertical upright *before* you build it?

You can use accurate construction to test ramp designs on paper

Accurate construction is different from making a sketch.

You'll need:

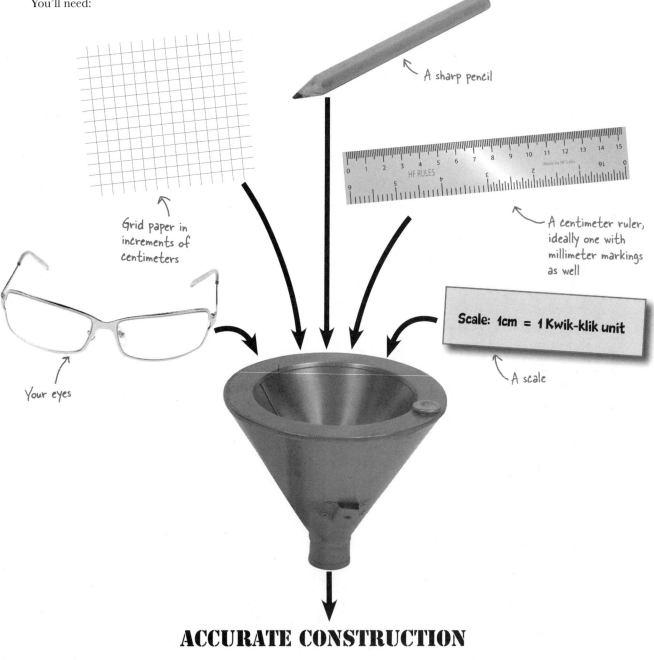

A sharp pencil

Grid paper in increments of centimeters

A centimeter ruler, ideally one with millimeter markings as well

HF RULES

Scale: 1cm = 1 Kwik-klik unit

A scale

Your eyes

ACCURATE CONSTRUCTION

Hello? Didn't you spend chapters 1 & 2 going on about how we couldn't trust a drawing, couldn't just measure it, blah, blah, blah?

True. But making an accurate construction is different.

Sharpen your pencil

How is making your own accurate drawing different from just measuring a sketch or diagram you're given? Write out your answer in words below.

Sharpen your pencil
Solution

How is making your own accurate drawing different from just measuring a sketch or diagram you're given?

If you're drawing the diagram yourself you can keep it in proportion, and you can get the angles right, too. You can use a set square or a protractor, or gridded paper, to make sure lines that are supposed to be perpendicular are drawn at right angles.

If you draw a line 3cm and another line 6cm—measured with a good ruler—you know for sure that the first line you drew is half the length of the other.

When you're given a sketch you don't know whether the person who drew it used a ruler and protractor to make the drawing accurate, or just did it roughly.

Your answer might be worded differently—that's OK, it's the thought that counts....

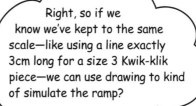

Right, so if we know we've kept to the same scale—like using a line exactly 3cm long for a size 3 Kwik-klik piece—we can use drawing to kind of simulate the ramp?

Yup. If you can't draw it with a right angle, then you can't build it with a right angle, either!

Centimeter grid paper has horizontal and vertical lines which are perfectly perpendicular—making it extra useful for drawing shapes with right angles like our ramps need to be. Just try it out....

Sharpen your pencil and ruler

Use accurate construction to find what length the vertical piece needs to be in order for the ramp to have a perpendicular upright.

Using a scale of 1cm to 1 Kwik-klik unit (so a piece of length 2 would be drawn as 2cm), use your ruler to find the part that fits.

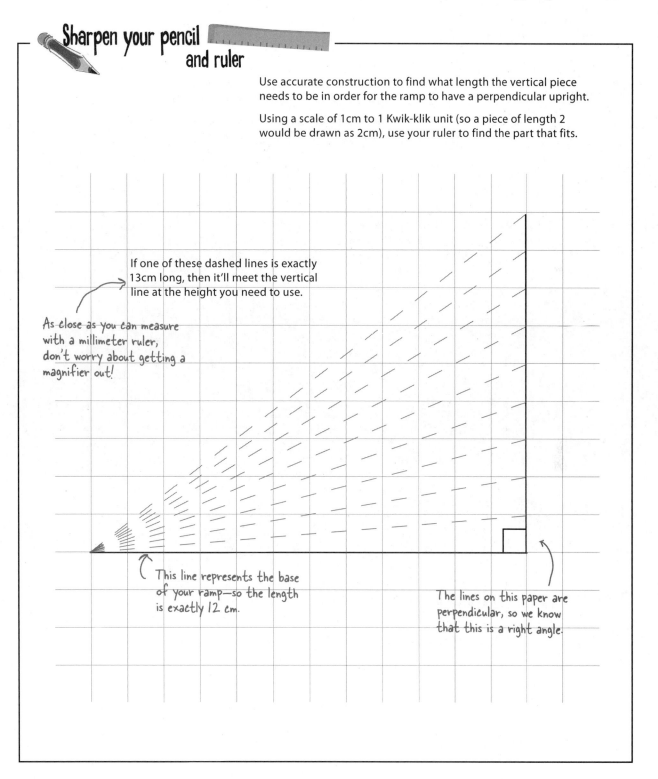

If one of these dashed lines is exactly 13cm long, then it'll meet the vertical line at the height you need to use.

As close as you can measure with a millimeter ruler, don't worry about getting a magnifier out!

This line represents the base of your ramp—so the length is exactly 12 cm.

The lines on this paper are perpendicular, so we know that this is a right angle.

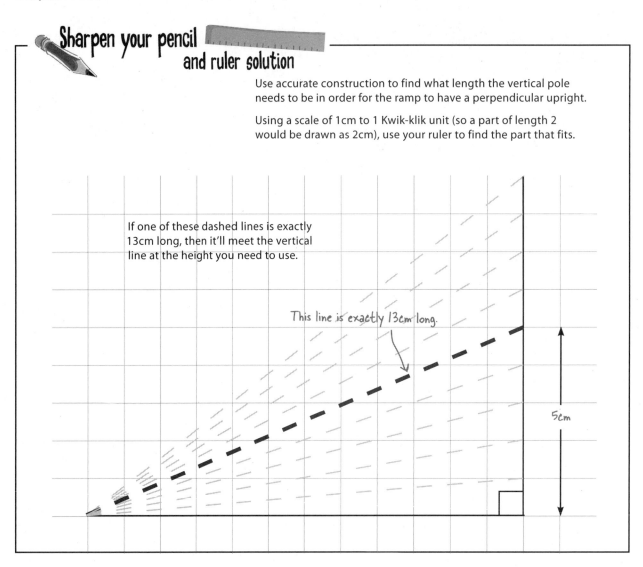

Sharpen your pencil and ruler solution

Use accurate construction to find what length the vertical pole needs to be in order for the ramp to have a perpendicular upright.

Using a scale of 1cm to 1 Kwik-klik unit (so a part of length 2 would be drawn as 2cm), use your ruler to find the part that fits.

If one of these dashed lines is exactly 13cm long, then it'll meet the vertical line at the height you need to use.

This line is exactly 13cm long.

5cm

The size 5 upright gives us a perfect vertical

The parts with lengths 12, 13, and 5 make a perfect ramp with a right angle between the horizontal and vertical parts.

Of course what we just drew was a scaled version of our final ramp—nobody wants a skate ramp 5cm high, but a ramp that's 5 kwik-klik units high is plenty big enough.

Sharpen your pencil
and ruler

Sam's sketched up another two jumps which she *hopes* can be built from the Kwik-klik parts.

Use the same pencil and ruler construction technique to find out whether there's a part that creates a perfect perpendicular ramp for each of them.

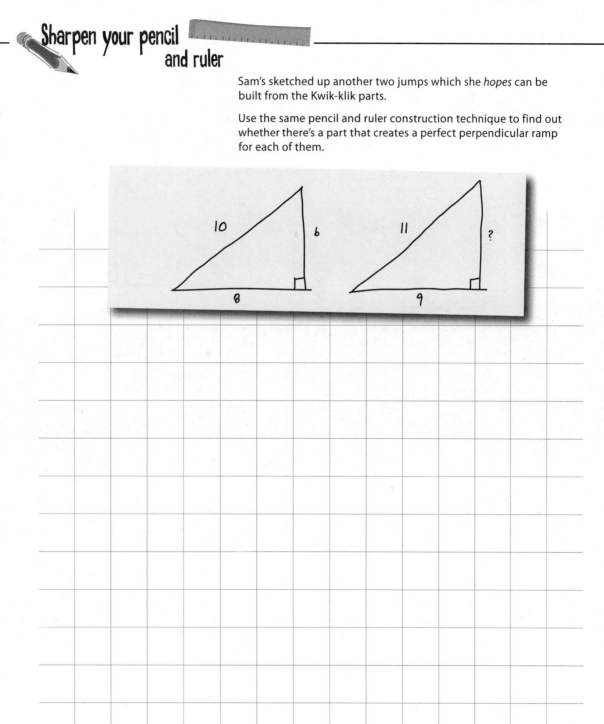

Sharpen your pencil and ruler solution

Sam's sketched up another two jumps which she *hopes* can be built from the Kwik-klik parts.

Use the same pencil and ruler construction technique to find out *whether* there's a part that creates a perfect perpendicular ramp for each of them.

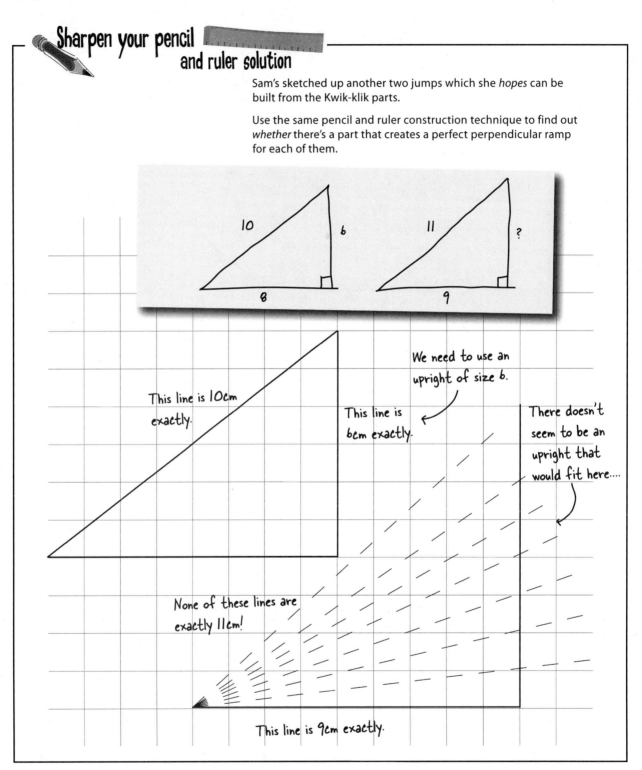

This line is 10cm exactly.

This line is 6cm exactly.

We need to use an upright of size 6.

There doesn't seem to be an upright that would fit here....

None of these lines are exactly 11cm!

This line is 9cm exactly.

Not all lengths make a right triangle

There isn't a length that can complete Sam's design for an 11 - 9 - ? jump and give a perfectly vertical upright. Size 7 is too long and size 6 is too short. It seems some lengths can make right triangles and others can't.

← Kwik-klik doesn't make a "six and a bit" size piece so you can only build jumps with whole number lengths.

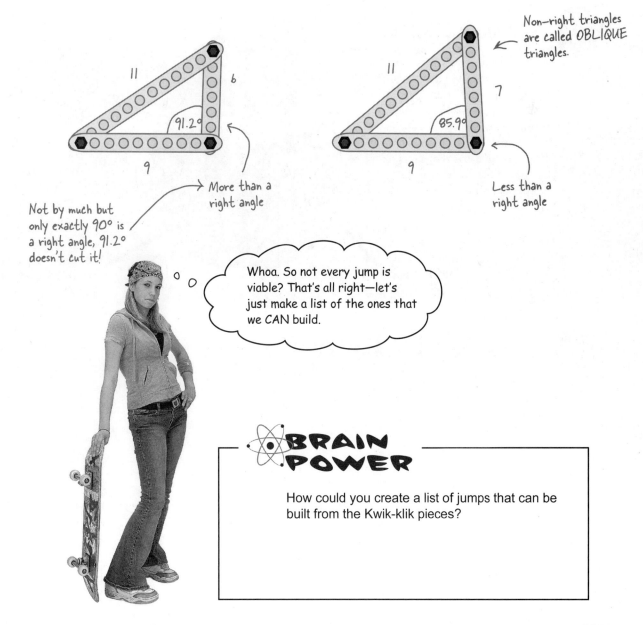

Non-right triangles are called OBLIQUE triangles.

11

6

91.2°

9

→ More than a right angle

Not by much but only exactly 90° is a right angle, 91.2° doesn't cut it!

11

7

85.9°

9

Less than a right angle

Whoa. So not every jump is viable? That's all right—let's just make a list of the ones that we CAN build.

BRAIN POWER

How could you create a list of jumps that can be built from the Kwik-klik pieces?

You can explore a geometry problem in different ways

There's more than one way to investigate a geometry problem, and the choice of which one to go for is often about how your own brain works rather than one being "better."

 Use your BRAIN

Your brain is amazing. Even before you started to learn about geometry, your brain could already recognize symmetry and special angles. But don't just think, use your imagination, too. What if this shape was *really* BIG? What if you turned it upside down?

Think about the RULES in your GEOMETRY TOOLBOX.

Use LOGIC to PROVE something works or doesn't work.

IMAGINE what would happen if....

 Use a PENCIL and PAPER

Sketches can help you think, but an accurate drawing can also show you whether something is possible or impossible.

Grid paper gives you a head start.

Add a ruler for more accurate drawings.

3 Use the REAL WORLD

Geometry isn't just important for your grades, it's what engineers and scientists rely on to make buildings stand up and cars go around corners.

Use kids' toys, a scrap piece of wood, or make models out of cardboard to put your theories to the test.

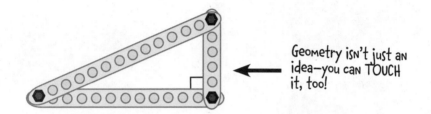

Geometry isn't just an idea—you can TOUCH it, too!

4 Use a COMPUTER MODEL

If you've ever played a video or computer game then you've already used a computer model of some geometry.

Whether the model is simple, or complex (like a racing game), the great thing is that you get to change something and then see what impact your change makes.

A computer model lets you test out stuff as if it was in the REAL WORLD.

So—what's your preference? Do you think it matters which exploration technique you use to find the jumps you can build?

In geometry, the rules are the rules

In geometry there's no "i before e *except* after c" stuff. The rules are the rules. Everything in your Geometry Toolbox applies whether you think, draw, touch, look, play, or test your way to an answer.

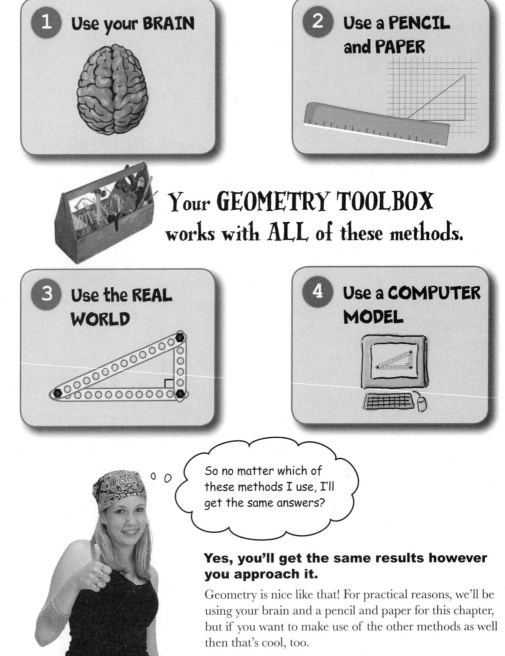

1 Use your BRAIN

2 Use a PENCIL and PAPER

Your **GEOMETRY TOOLBOX** works with **ALL** of these methods.

3 Use the REAL WORLD

4 Use a COMPUTER MODEL

So no matter which of these methods I use, I'll get the same answers?

Yes, you'll get the same results however you approach it.

Geometry is nice like that! For practical reasons, we'll be using your brain and a pencil and paper for this chapter, but if you want to make use of the other methods as well then that's cool, too.

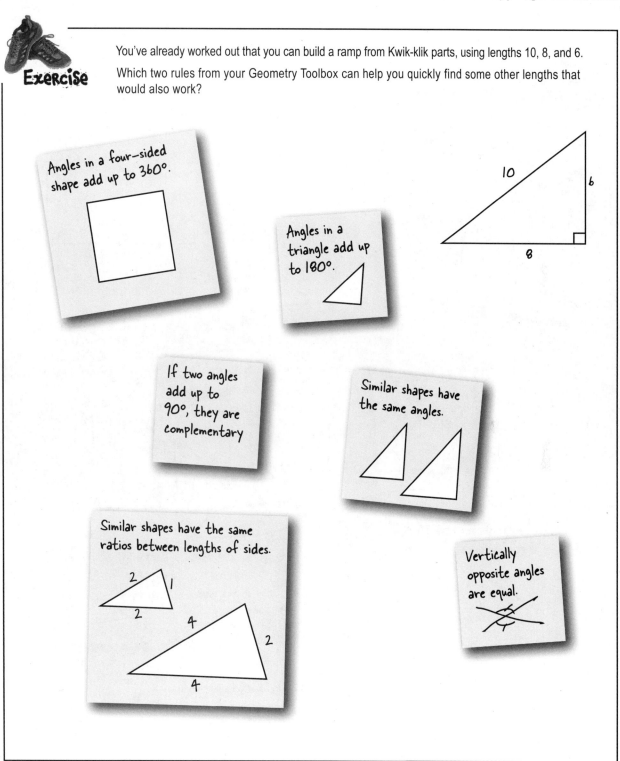

Exercise

You've already worked out that you can build a ramp from Kwik-klik parts, using lengths 10, 8, and 6.

Which two rules from your Geometry Toolbox can help you quickly find some other lengths that would also work?

Angles in a four-sided shape add up to 360°.

Angles in a triangle add up to 180°.

10   6   8

If two angles add up to 90°, they are complementary

Similar shapes have the same angles.

Similar shapes have the same ratios between lengths of sides.

2   1   2

4   2   4

Vertically opposite angles are equal.

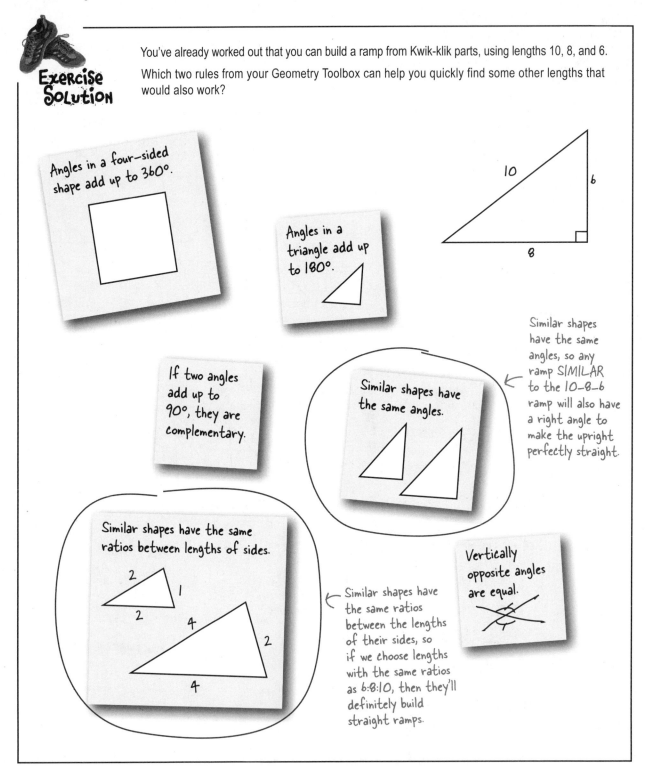

Exercise Solution

You've already worked out that you can build a ramp from Kwik-klik parts, using lengths 10, 8, and 6.

Which two rules from your Geometry Toolbox can help you quickly find some other lengths that would also work?

Angles in a four-sided shape add up to 360°.

Angles in a triangle add up to 180°.

10   6

8

If two angles add up to 90°, they are complementary.

Similar shapes have the same angles.

Similar shapes have the same angles, so any ramp SIMILAR to the 10–8–6 ramp will also have a right angle to make the upright perfectly straight.

Similar shapes have the same ratios between lengths of sides.

2   1
2

4   2

4

Similar shapes have the same ratios between the lengths of their sides, so if we choose lengths with the same ratios as 6:8:10, then they'll definitely build straight ramps.

Vertically opposite angles are equal.

Any good jump has some similar scaled cousins

When you scale the ramp—making it bigger or smaller—none of the angles change, so it stays a good ramp with a right angle.

To find whole number lengths for the smaller-scaled sizes, look for a common factor in your current lengths. That's easiest to do by using a set of factor trees.

Use factor trees to find whole-number miniatures

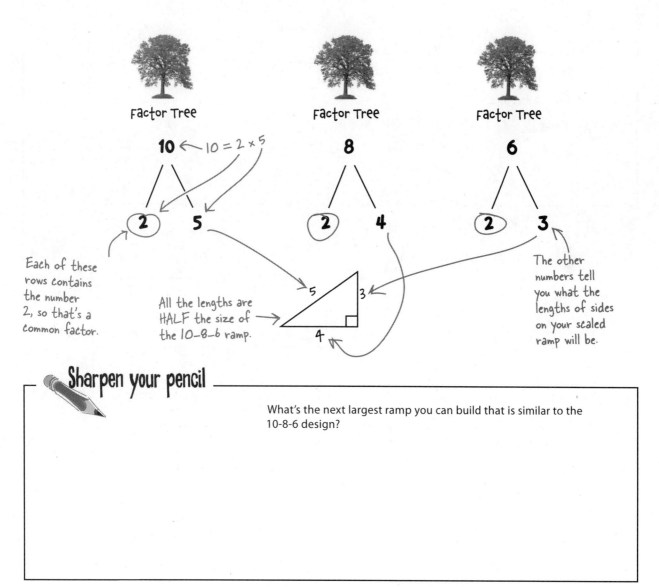

Factor Tree   Factor Tree   Factor Tree

10 ← 10 = 2 × 5   **8**   **6**

2   5   2   4   2   3

Each of these rows contains the number 2, so that's a common factor.

All the lengths are HALF the size of the 10–8–6 ramp. →

5   3
4

The other numbers tell you what the lengths of sides on your scaled ramp will be.

Sharpen your pencil

What's the next largest ramp you can build that is similar to the 10-8-6 design?

Sharpen your pencil
Solution

What's the next largest ramp you can build that is similar to the 10-8-6 design?

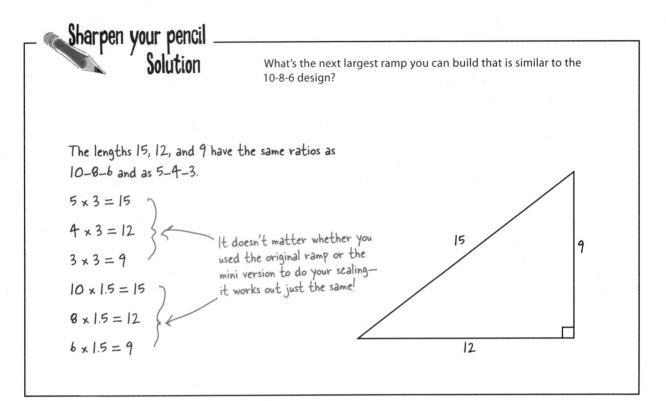

The lengths 15, 12, and 9 have the same ratios as 10–8–6 and as 5–4–3.

$5 \times 3 = 15$

$4 \times 3 = 12$

$3 \times 3 = 9$

$10 \times 1.5 = 15$

$8 \times 1.5 = 12$

$6 \times 1.5 = 9$

It doesn't matter whether you used the original ramp or the mini version to do your scaling— it works out just the same!

15

9

12

---

there are no
Dumb Questions

Q: What if I wanted a ramp even smaller? Can I just keep doing more factor trees?

A: The Kwik-klik units don't come in half sizes—there isn't a 1.5 length, so you'd quickly run out of parts, but assuming you weren't just talking about building it with the kit parts, you still only need to do your factor trees until one of the numbers on the bottom is a prime number—that means it can't be divided by anything except itself and one. Then to make a really small ramp you'd multiply those factors by a fraction.

Q: How can we skate on a 2D ramp? Isn't this gonna be more like a rail that you can slide on?

A: What we're actually representing is the side of the ramp. There would be two of these side triangles the same, with a panel connected to the sloping beam on each triangle. This is a 3D problem which has a 2D solution.

If you're interested in exploring 3D problems further, come and catch up with Sam in *Head First 3D Geometry*.

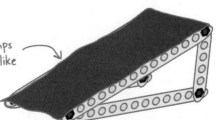

What the jumps actually look like in 3D

So how do we know which sets of ratios can give us a right triangle? The dude at the store gave me this slip of paper with some odd stuff on it—tips for building right angles or something.... I didn't really pay it any attention—what do you think it means?

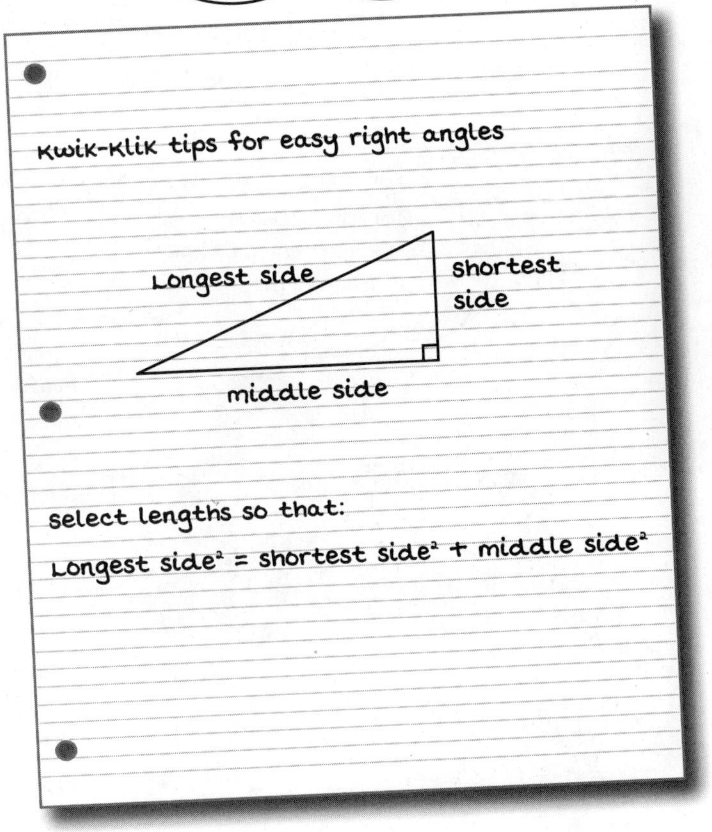

Kwik-Klik tips for easy right angles

Longest side / shortest side

middle side

select lengths so that:

Longest side^2 = shortest side^2 + middle side^2

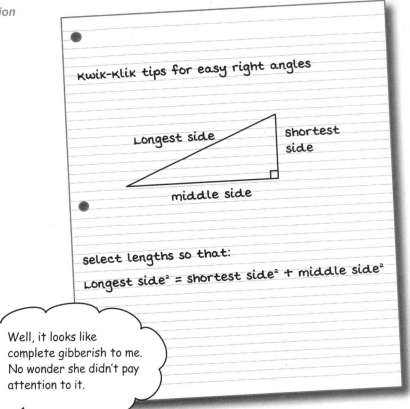

Kwik-Klik tips for easy right angles

Longest side

shortest side

middle side

select lengths so that:

Longest side² = shortest side² + middle side²

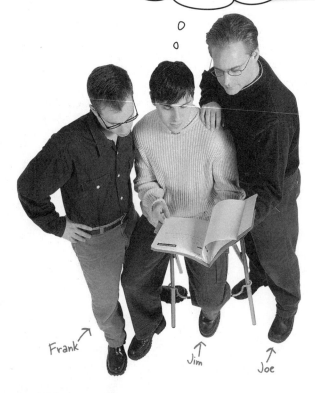

Well, it looks like complete gibberish to me. No wonder she didn't pay attention to it.

Frank ↗

Jim ↑

Joe ↑

Frank: But it must mean something. I mean, nobody goes to the trouble of writing something down unless it's useful.

Jim: True. But how would you use it? And why would squaring the side lengths have anything to do with the angle of the triangle?

Joe: Oh…those twos are for squaring! Yeah—no way that would work.

Frank: OK, don't freak out, but if you just try it, like for the 3-4-5 ramp design…it works out perfect.

Jim: What? Are you sure you got your numbers right?

Frank: Yeah—I'm sure. Longest side is 5, and 5 squared is 25. Shortest side is 3, and 3 squared is 9, and the middle side is 4, and 4 squared is 16. So—add up the middle and shortest sides squared—9 plus 16—and you get….

Joe: 25. The same as the square of the longest side. That has got to be a coincidence.

Frank: There's only one way to find out—let's check the others.…

Geometry Detective

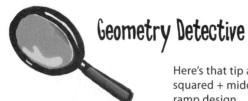

Here's that tip again: longest-side squared = shortest-side squared + middle-side squared seems to work for at least one ramp design.

Based on the four jumps you've successfully designed so far, complete the table to discover whether there really is a secret pattern behind designing perfect ramps every time.

Triangle	3-4-5	6-8-10	5-12-13	9-12-15
Shortest-side's length	3			
Middle-side's length	4			
Longest-side's length	5			
Shortest-side squared	9			
Middle-side squared	16			
Longest-side squared	25			
Shortest-side squared + longest-side squared	9 + 16 = 25			

Add these two together

Now, isn't THAT kind of freaky....

The lengths of the sides are linked by a pattern

Did you find it? For the four right triangles we tested, it seems like the square of the length of the longest side is equal to the squares of the other two sides added together.

Triangle	3-4-5	6-8-10	5-12-13	9-12-15
Shortest-side's length	3	6	5	9
Middle-side's length	4	8	12	12
Longest-side's length	5	10	13	15
Shortest-side squared	9	36	25	81
Middle-side squared	16	64	144	144
Longest-side squared	25	100	169	225
Shortest-side squared + middle-side squared	9 + 16 = 25	36 + 64 = 100	25 + 144 = 169	81 + 144 = 225

This row
+
this row
= this row!

> Right. So—for these four right triangles, you get this freaky pattern. Don't you think if you're gonna use this to design skate jumps you need something a bit more reliable? What if these are the ONLY four triangles it works for? It doesn't even make sense. What have squares got to do with triangles anyway?

True. Can we really trust this pattern?

The numbers we've tested so far seem pretty conclusive, but do we definitely know that this pattern is going to work for all the possible right-angled jumps we might need?

And how do the squares relate to the triangles? Let's investigate this pattern in a more general way.

Geometry Investigation Magnets

Let's experiment with a general right triangle. The sides of the triangle can be a, b, and c—where c is the longest side. Below are two large squares, each of which has side length a+b.

Can you arrange the gray triangles inside the squares so that in one box you are left with a square with side length "c" and in the other box you are left with two squares—one with side length "a" and one with side length "b"? Make sure to use four triangles in each box.

What do you know about the white area left in each box? What does this tell you about how the pattern you found might work for a right triangle with sides a, b, and c?

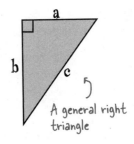

A general right triangle

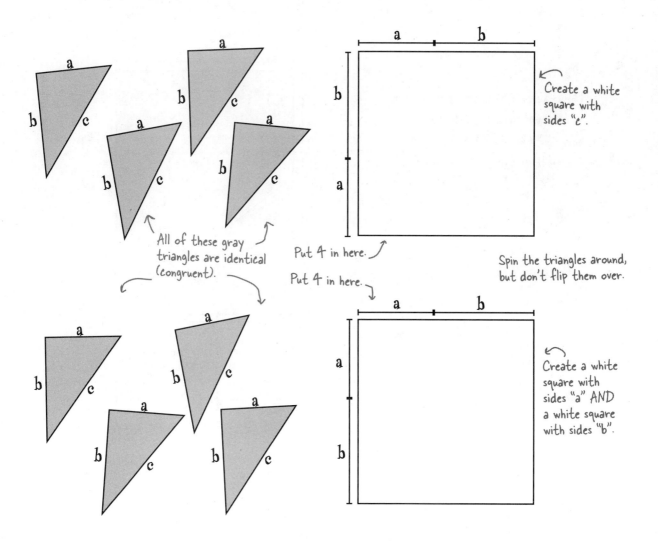

All of these gray triangles are identical (congruent).

Put 4 in here.

Put 4 in here.

Create a white square with sides "c".

Spin the triangles around, but don't flip them over.

Create a white square with sides "a" AND a white square with sides "b".

Geometry Investigation Magnets Solution

Let's experiment with a general right triangle. The sides of the triangle can be a, b, and c—where c is the longest side. Below are two large squares, each of which has side length a+b.

Can you arrange the gray triangles inside the squares so that in one box you are left with a square with side length "c" and in the other box you are left with two squares—one with side length "a" and one with side length "b"? Make sure to use four triangles in each box.

What do you know about the white area left in each box? What does this tell you about how the pattern you found might work for a right triangle with sides a, b, and c?

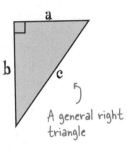

A general right triangle

The gray triangles are all congruent, so the gray area we've created in each box must be equal. This means that the leftover area in white must also be equal.

For a right triangle with sides a, b, and c, the square of c is equal to the squares of a and b added together.

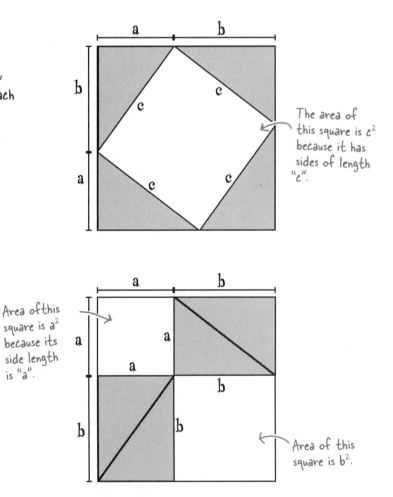

The area of this square is c^2 because it has sides of length "c".

Area of this square is a^2 because its side length is "a".

Area of this square is b^2.

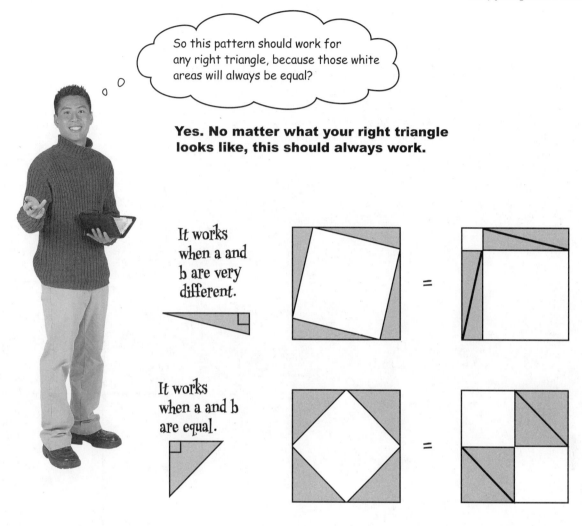

So this pattern should work for any right triangle, because those white areas will always be equal?

Yes. No matter what your right triangle looks like, this should always work.

It works when a and b are very different.

It works when a and b are equal.

BRAIN BARBELL

You've got a skate course to design! Drawing triangles and squares might be a good way to investigate the pattern, but it's probably not the most useful way to capture it for reuse. How could you capture this pattern:

1. As words? ....

....

2. Using algebra (use the letters a, b, and c like in the magnets exercise)

....

The square of the longest side is equal to the squares of the other two sides added together

Your answer might be in words or letters.

This has to be one of the most amazing patterns you'll discover in geometry. Whenever you have a right triangle, if you draw a square on the longest side, its area is *exactly* equal to the squares you could draw on the other two sides.

Area = 100

36 + 64 = 100

10

10×10

6   Area = 36

8

6×6

This works for EVERY right triangle you could draw.

Area = 64

8×8

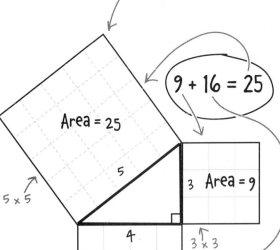

Area = 25

9 + 16 = 25

5

5×5

3   Area = 9

4

3×3

Area = 16

4×4

The Pythagorean Theorem: a² + b² = c²

This pattern is known as ***The Pythagorean Theorem***. Using it you can find out whether the corner opposite the longest side of a triangle is acute, obtuse, or a right angle. In words, the theorem is usually written:

← Author's note: In British English, it's known as Pythagoras' Theorem.

The sum of the squares of the legs of a right triangle is equal to the square of the hypotenuse.

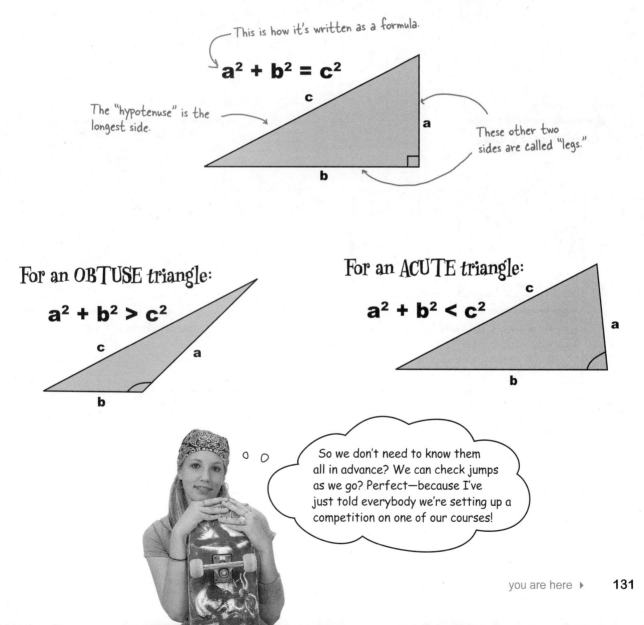

This is how it's written as a formula.

$$a^2 + b^2 = c^2$$

The "hypotenuse" is the longest side.

c

a

b

These other two sides are called "legs."

For an OBTUSE triangle:

$$a^2 + b^2 > c^2$$

c

a

b

For an ACUTE triangle:

$$a^2 + b^2 < c^2$$

c

a

b

So we don't need to know them all in advance? We can check jumps as we go? Perfect—because I've just told everybody we're setting up a competition on one of our courses!

Skate Ramp Magnets

Sam's sketched out a rough design for the competition course. She wants three different sets of ramps for the skaters to demo their skills on—one with a gap to jump and two with joining rails to slide on.

Using the Pythagorean Theorem and the construction kit magnets, work out where each part needs to go to complete the course design. Use each part exactly once.

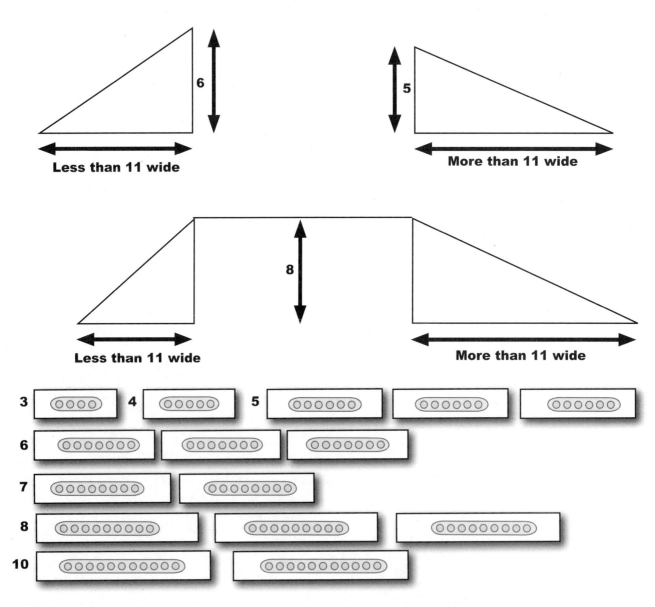

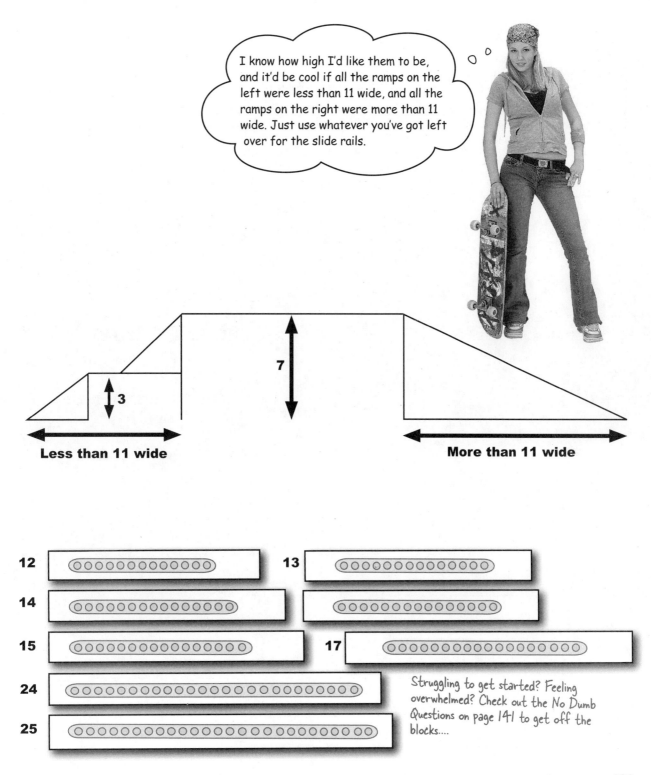

I know how high I'd like them to be, and it'd be cool if all the ramps on the left were less than 11 wide, and all the ramps on the right were more than 11 wide. Just use whatever you've got left over for the slide rails.

7

3

Less than 11 wide

More than 11 wide

12

13

14

15   17

24

25

Struggling to get started? Feeling overwhelmed? Check out the *No Dumb Questions* on page 141 to get off the blocks....

Skate Ramp Magnets Solution

Sam's sketched out a rough design for the competition course. She wants three different sets of ramps for the skaters to demo their skills on—one with a gap to jump and two with joining rails to slide on.

Using the Pythagorean Theorem and the construction kit magnets, work out where each part needs to go to complete the course design. Use each part exactly once.

Solving this problem is mostly a matter of trial and error—checking to see where you can find lengths which fit the Pythagorean Theorem: Longest side squared = sum of other two sides squared.

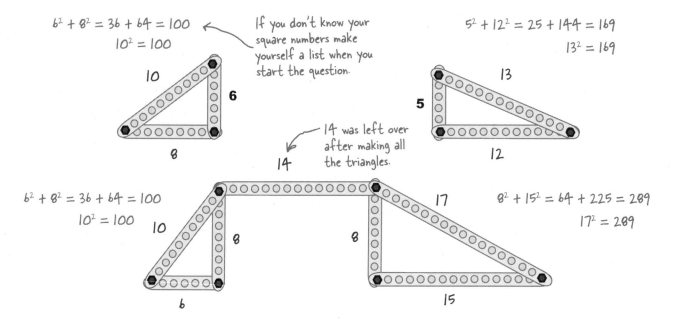

$6^2 + 8^2 = 36 + 64 = 100$
$10^2 = 100$

If you don't know your square numbers make yourself a list when you start the question.

10   6   8

$5^2 + 12^2 = 25 + 144 = 169$
$13^2 = 169$

13   5   12

14 was left over after making all the triangles.

14

$6^2 + 8^2 = 36 + 64 = 100$
$10^2 = 100$

10   8   8   6

17   15

$8^2 + 15^2 = 64 + 225 = 289$
$17^2 = 289$

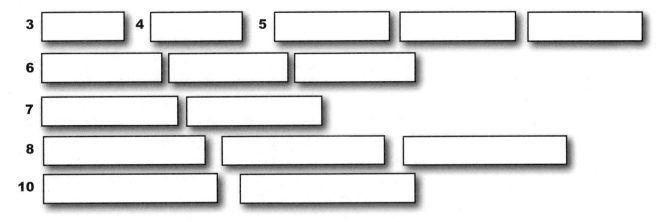

3   4   5

6

7

8

10

Don't worry if you didn't do it this way—but it's a good tip to get yourself going.

The key to finding sets of lengths that work for right triangles is to remember that the Pythagorean pattern is all about combining square numbers.

So—you could start by writing down each length, and its square, and then look to see whether the difference between any of your square numbers is the same as another square number you've written down.

When you've got one side length you even know which difference you're looking for: e.g., short side 8 = difference 64.

Lengths	Squares
3	9
4	16
5	25
6	36
7	49
8	64
10	100
12	144
13	169
14	196
15	225
17	289
24	576
25	625

Difference between 576 and 625 is 49

This part here is tricky—you might have recognized that you need to use two of those useful 3–4–5 triangles.

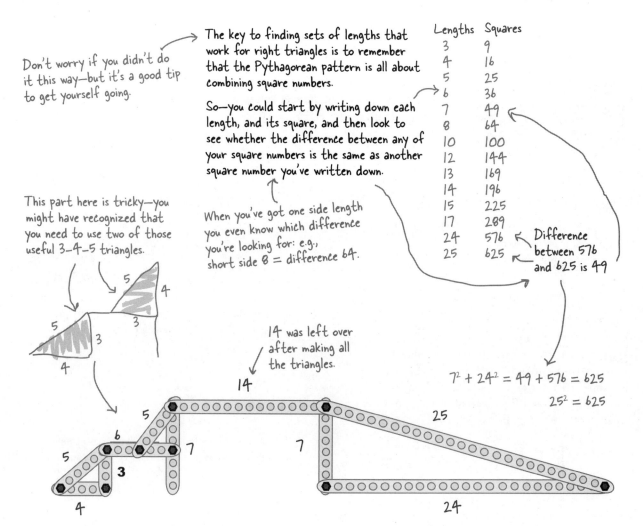

14 was left over after making all the triangles.

$7^2 + 24^2 = 49 + 576 = 625$

$25^2 = 625$

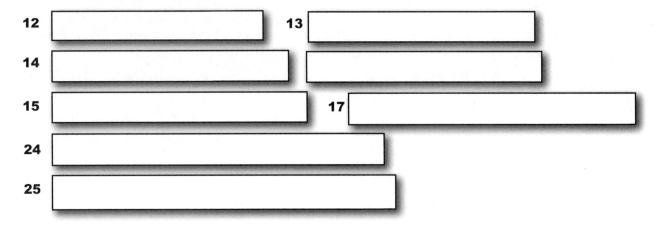

12 □ 13 □
14 □ □
15 □ 17 □
24 □
25 □

Fireside Chats

Tonight's talk: **Integer solutions vs non-integer solutions for right triangles**

Integer solutions:

I guess we're quite a famous family. Most people can recognize at least one or two of us. 3-4-5 gets the most attention but that's not the whole range.

Yeah but you totally need a calculator—nobody knows the square roots of weird numbers by heart… and who carries one of those around? We integer solutions are convenient—especially if you can remember them!

Us—precious? You're the one with all those high-maintenance decimal points, baby.…

That's true. I always think you look better that way actually.…

Non-integer solutions:

Range schmange. We don't have a range—think of a number, and we can find you a solution. We non-integer solutions are flexible. Any size, any place, you can still have a right angle.

Remember? People forget we even exist—never mind learning specific versions of us…you lot are so precious.

Oh, I knew you'd bring that up. Yeah, yeah, we're detailed—OK? But cell phones and computers have calculators, and then you don't have to remember a thing—just use us where you need. And don't forget you can always write a root as a root.…

Using Kwik-klik skate ramps is definitely the right angle!

By finding right triangles with whole number ratios between the lengths of the sides, you've managed to build a whole load of different ramps from the Kwik-klik standard parts. It's a fast, cheap, and smart way to build a skate course, and you're in a great position to reap the rewards.

That was so easy! I can't wait for the next one. Do you want your cut of the money now, or should I just let it pile up for a while?

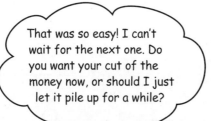

SKATE UNLIMTED BLOG

http://skateunlimetedblog.com

02 JULY

Skate Comp success

Unbelievable! Not only did the course get assembled in a day but man, it was the hottest street course we've seen in a long time. Some serious kick-flipping action was had over some huge jumps, and we just about tail-slid our way to heaven on the excellent ramp-and-rail section.

Fran 'five-oh' Sampson took the overall number one spot at the end of the day, but we were all grinning like winners thanks to Sam & Co's fantastic design-and-build efforts.

> click for pics

BULLET POINTS

- To find the longest side, square the other two sides, add those up, and then find the square root.

- To find a short side, square the other two sides and subtract one from the other. Make it a positive number and then find the square root.

- Finding missing sides using the Pythagorean Theorem only works for right triangles.

there are no Dumb Questions

Q: But WHY? How come it adds up like that?

A: That is a very good question and one which mathematicians and philosophers have struggled with for centuries. Unfortunately there's not always good "why" reasons in geometry—stuff just IS. The good news is that it's very reliable…so while understanding why the Pythagorean Theorem works is beyond most of us, at least we can make good use of it.

Q: Why is it called The Pythagorean Theorem? Why not something more meaningful and easier to spell like "Right triangle check" theorem?

A: Pythagoras was a Greek guy. He was first to write down the theorem. That isn't to say that nobody had noticed it before, but he gets the props. Shame he wasn't called something easier to spell!

Q: I tried to use the Pythagorean Theorem to find a missing short side, and I was stuck trying to find the square root of a negative number. What should I do?

A: You've probably got your legs and your hypotenuse mixed up. Try to redo your subtraction the other way round. This should give you a positive number, and you'll be able to find the square root.

Q: So is "c" always the hypotenuse?

A: The Pythagorean Theorem uses "c" for the hypotenuse and "a" and "b" for the legs. Of course we know that algebra is just a tool for describing a pattern, and in this case, the pattern is what's important: **the longest side squared equals the other two sides squared.** So, if you get mixed up about your a, b, and c, think about what the pattern behind that formula is, and go from there.

Q: Do all triangles have a hypotenuse? What if I have two long sides of equal length and one short side? Which is the hypotenuse?

A: Only right triangles officially have a hypotenuse. And if you've got two sides equally long, and one shorter one, your triangle can't have a right angle, because it would fail the $c^2 = a^2 + b^2$ test, whichever of your longer sides you decided was "c". But don't think you can't use the Pythagorean Theorem for triangles that don't have a right angle…it's just takes a little more thought to apply it. (More on that in a minute.)

OMG! Come read this email—this is so cool… you won't believe it!

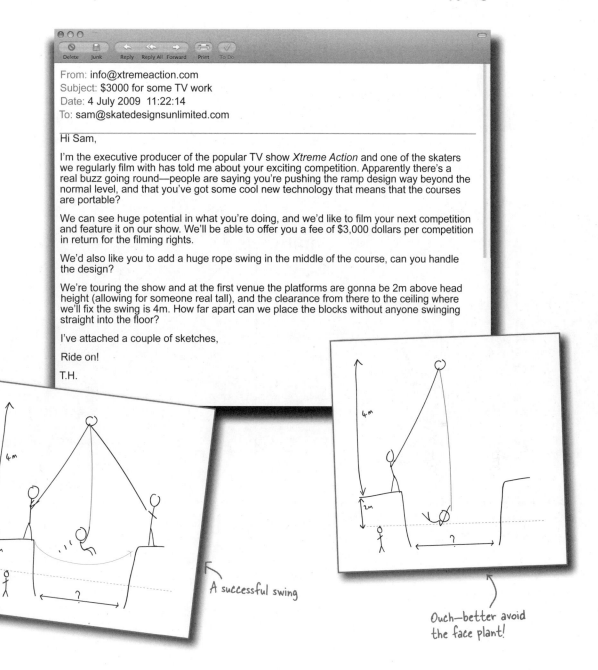

From: info@xtremeaction.com
Subject: $3000 for some TV work
Date: 4 July 2009 11:22:14
To: sam@skatedesignsunlimited.com

Hi Sam,

I'm the executive producer of the popular TV show *Xtreme Action* and one of the skaters we regularly film with has told me about your exciting competition. Apparently there's a real buzz going round—people are saying you're pushing the ramp design way beyond the normal level, and that you've got some cool new technology that means that the courses are portable?

We can see huge potential in what you're doing, and we'd like to film your next competition and feature it on our show. We'll be able to offer you a fee of $3,000 dollars per competition in return for the filming rights.

We'd also like you to add a huge rope swing in the middle of the course, can you handle the design?

We're touring the show and at the first venue the platforms are gonna be 2m above head height (allowing for someone real tall), and the clearance from there to the ceiling where we'll fix the swing is 4m. How far apart can we place the blocks without anyone swinging straight into the floor?

I've attached a couple of sketches,

Ride on!

T.H.

A successful swing

Ouch—better avoid the face plant!

The TV company wants you to design them a rope swing, so that they get maximum excitement and minimum lawsuits. Are they right to be worried about accidents if the blocks at either side of the swing are too far apart?

A longer rope swings further and <u>lower</u>

The wider the gap between the platforms, the more exciting the swing
will be—but there's a catch. The TV company doesn't just want the
widest swing, they want the widest possible swing without people
smashing into the floor!

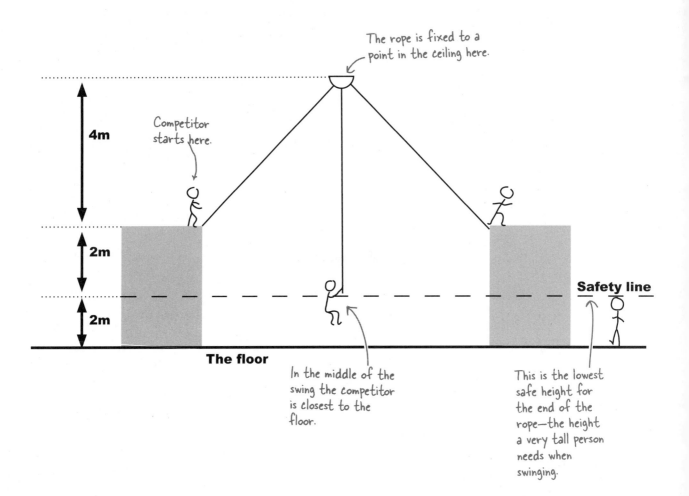

The rope is fixed to a
point in the ceiling here.

4m

Competitor
starts here.

2m

2m

The floor

In the middle of the
swing the competitor
is closest to the
floor.

Safety line

This is the lowest
safe height for
the end of the
rope—the height
a very tall person
needs when
swinging.

Sharpen your pencil

What's the longest the rope can be without going below the safety line?

there are no Dumb Questions

Q: The Pythagorean Theorem formula ($c^2 = a^2 + b^2$) looks like it's for finding the hypotenuse (c). But sometimes we're finding a length of a short side. How do I know what I'm supposed to be finding and how to find it?

A: You can rearrange the formula to focus on one of the short sides (legs). Like $a^2 = c^2 - b^2$ or $b^2 = c^2 - a^2$, but the most reliable thing to do is to focus on the meaning behind the pattern. If you're looking for the hypotenuse remember that the squares of the two shortest sides add up to make the square of the longest side.

If you're finding a short side then you need to think of the pattern like this: the difference between the square of the longest side and another side is the square of the remaining side.

If you're stuck trying to design the skate course on page 133, then try this to get you started.

Q: OK, that's not so bad when I'm given the lengths of two sides and I have to find the other side's length. But what about when I'm only given one side and I have to find integer values that complete a right triangle? The formula can't give me the answer because I don't have enough values!

A: Again, think about the pattern behind the formula. If you've got a load of possible values—or if you know the value is within a range (like "less than 20")—then here's a trick you can rely on to find the answers: just remember that this pattern is about square numbers. (1, 4, 9, 16, 25, 36, 49, 64, 81, 100,...etc).

Write down all the values you've been given to choose between, and then write down their squares. Then compare pairs of numbers and see whether they add up to a square number, or whether the difference between them is a square number.

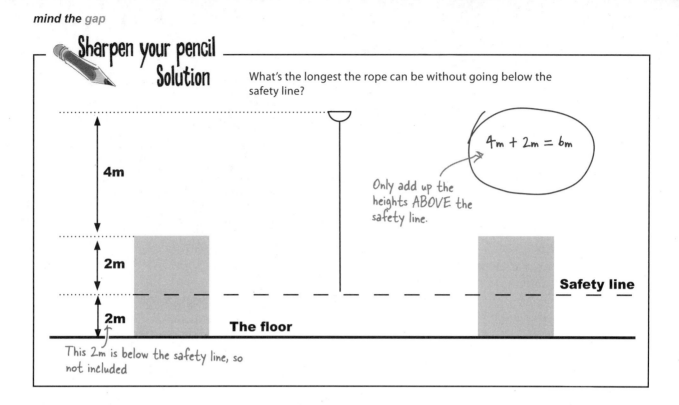

Sharpen your pencil
Solution

What's the longest the rope can be without going below the safety line?

4m

2m

2m

The floor

Safety line

Only add up the heights ABOVE the safety line.

4m + 2m = 6m

This 2m is below the safety line, so not included

So, how far can you swing on a six-meter rope?

The gap between the platforms is the base of a triangle, with the rope making up two of the sides joined at the point where the rope is fixed at the top. So the distance of the gap is the same as the length of the base of the triangle.

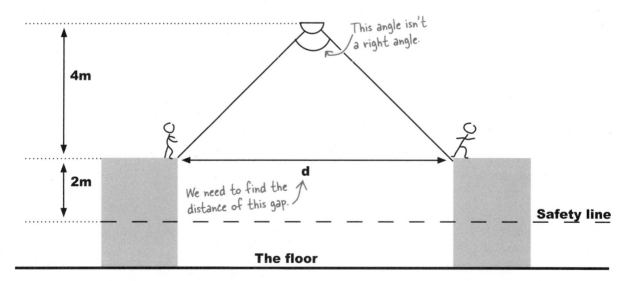

4m

2m

This angle isn't a right angle.

d

We need to find the distance of this gap.

Safety line

The floor

Well, the rope doesn't change length, does it? So I guess that gives us two sides of the triangle, but we can't use the Pythagorean Theorem to find the missing side without a right angle, can we?

That's right—the Pythagorean Theorem only finds missing sides for right triangles.

So, when you're faced with a triangle without a right angle you've got two options: find something else in your Geometry Toolbox to solve the problem, or see if you can somehow turn your non-right triangle into a right triangle (or triangles!)

BRAIN BARBELL

How can you make the swing problem into a right triangle problem so that you can use the Pythagorean Theorem to find length d?

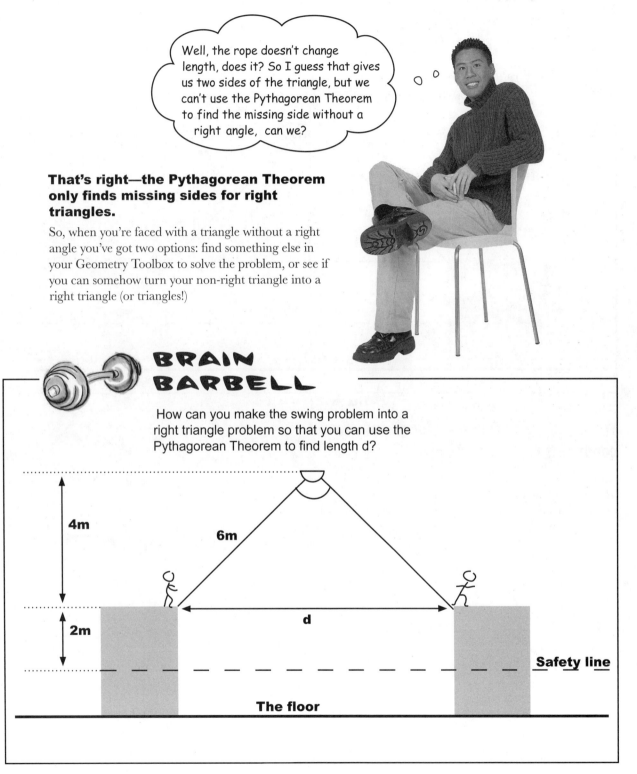

4m

6m

d

2m

Safety line

The floor

BRAIN BARBELL SOLUTION

How can you make the swing problem into a right triangle problem so that you can use the Pythagorean Theorem to find d?

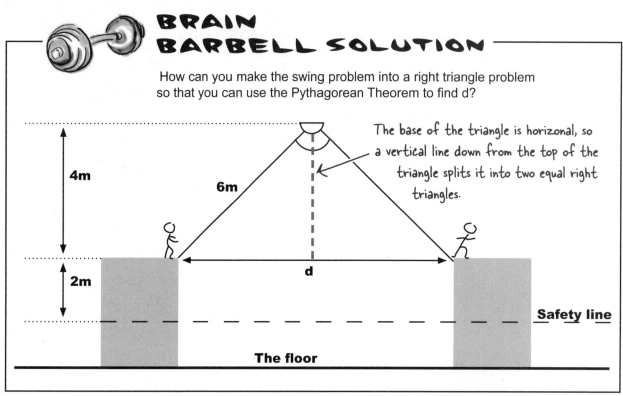

The base of the triangle is horizonal, so a vertical line down from the top of the triangle splits it into two equal right triangles.

You can split an isoceles triangle into two congruent right triangles

You can split a triangle into two right triangles by drawing an **altitude**—a line which joins the top of the triangle to the base, perpendicular to the base. An isoceles triangle has two sides and two angles the same, so the two triangles created are congruent.

Bisect

Chop into two equal parts.

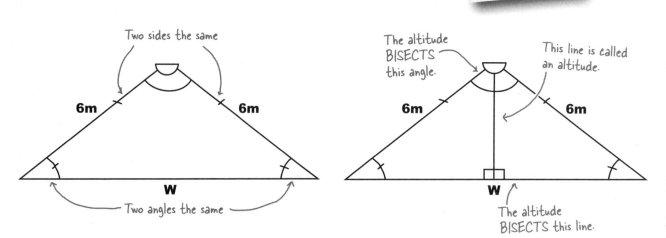

BE the Rope Swing

Your job is to play like you're the 6m rope. Use the Pythagorean Theorem to work out how far across a gap someone could swing on you.

---
there are no
Dumb Questions

Q: I get that the rope length doesn't change, so the sides are equal, but how did we know that the bottom two angles are equal?

A: An isoceles triangle has two sides equal but also two angles equal—always. So—if you see that the sides are the same you know the angles are the same, and the other way around.

Q: How did we know that the altitude would be vertical?

A: The altitude is always perpendicular to the base that it's drawn on, so if that base is horizontal (as it is in this case), then the altitude must be vertical.

Q: So, which triangle does the altitude belong to?

A: Both! The altitude is the shared side of the two identical (congruent) triangles it creates in this case. So, it belongs to both of them.

BE the Rope Swing Solution

Your job is to play like you're the 6m rope. Use the Pythagorean Theorem to work out how far across a gap someone could swing on you.

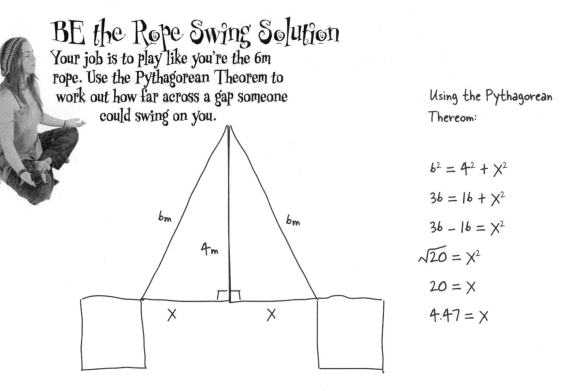

Using the Pythagorean Thereom:

$$6^2 = 4^2 + X^2$$
$$36 = 16 + X^2$$
$$36 - 16 = X^2$$
$$\sqrt{20} = X^2$$
$$20 = X$$
$$4.47 = X$$

Total distance you can swing = $2X = 2 \times 4.47 = 8.94$ meters

there are no Dumb Questions

Q: Wait a second—weren't we looking for integer solutions? How come 8.94 meters is OK?

A: The Kwik-klik parts only came in integer lengths, so we needed integer solutions for our ramps, but for the rope swing a non-integer solution is fine. We can put blocks 8 meters and 94 centimeters apart.

Q: Right. So, generally, is this Pythagorean thing for finding integer solutions or not?

A: You'll find that a lot of geometry problems center around the "Pythagorean triples." You can think of them as *Super Triangles*—right triangles with integer side lengths. You've already found some—3,4,5 and 5,12,13 are two of the most important.

Q: Super Triangles...cool! Do I need to learn the actual values or just know that they exist?

A: You can always use the Pythagorean Theorem to find them, but that's gonna be pretty slow...so if you're good at remembering stuff then learning the first three or four (in the toolbox at the end of this chapter) can be a real time saver.

Your rope swing is perfect

The design is exactly right, so your swing goes straight into action on the first show of the series.

On tonight's XTreme Show we bring you the action from the amazing new portable street course that's sweeping the nation, and we push gap jumping to the limits with the first ever mid-course rope swing!

Sweet! I've got tickets for both of us to travel around with the show for the whole season, plus our first pay check. Hmmm... a few new skateboards, and hello Apple store!

Your Geometry Toolbox

You've got Chapter 3 under your belt and now you've added the Pythagorean Theorem to your tool box. For a complete list of tooltips in the book, head over to www.headfirstlabs.com/geometry.

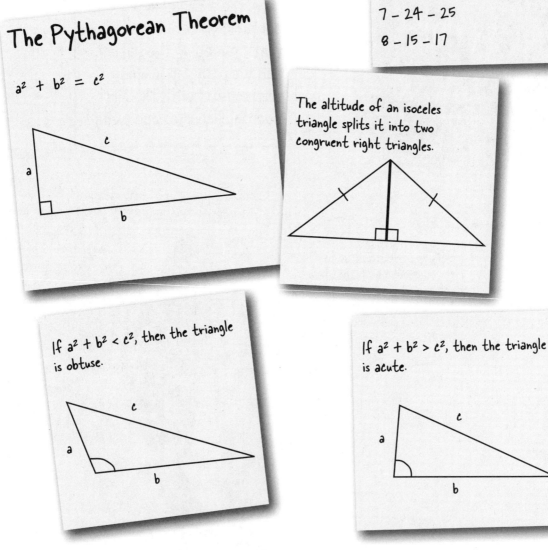

Super Triangles

(Also known as Pythagorean triples)

3 – 4 – 5

5 – 12 – 13

7 – 24 – 25

8 – 15 – 17

The Pythagorean Theorem

$a^2 + b^2 = c^2$

The altitude of an isoceles triangle splits it into two congruent right triangles.

If $a^2 + b^2 < c^2$, then the triangle is obtuse.

If $a^2 + b^2 > c^2$, then the triangle is acute.

4 triangle properties

Between a rock and show
a ~~hard~~ place
triangular

> Great news, Doreen—I just got a bigger corner office! At least, I think it's bigger...there's lots of windows. Do you have a tape measure handy?

Ever had that sinking feeling that you've made a bad decision?

In the real world, *choices can be complex*, and wrong decisions can cost you **money** and **time**. Many solutions aren't always straightforward: even in geometry, bigger doesn't always mean better—it might not even mean longer. *So what should you do?* The good news is that you can *combine your triangle tools* to **make great decisions** even when it seems like you don't have the right information to answer the question.

Everybody loves organizing a rock festival

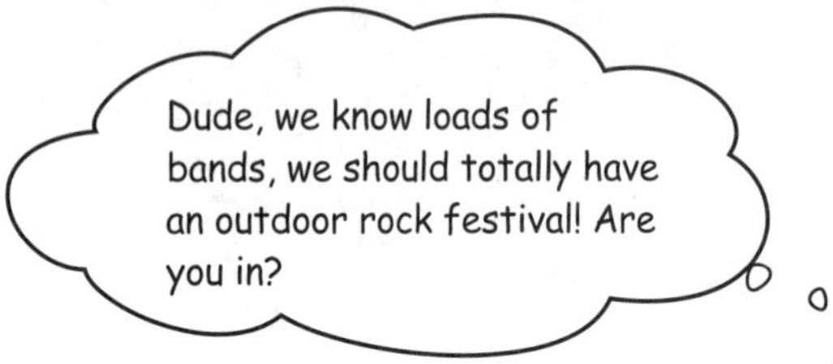

> Dude, we know loads of bands, we should totally have an outdoor rock festival! Are you in?

Your buddies have big plans for a local rock festival, but these things don't just organize themselves. There's a venue to choose, security fencing to sort out, a sound system to select....

Think you can handle it?

Your buddies

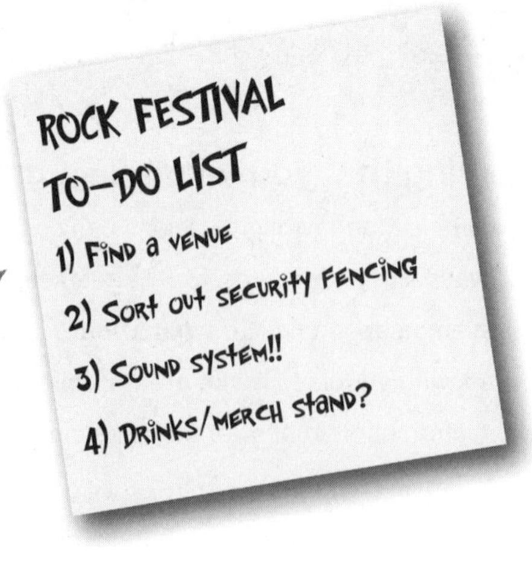

ROCK FESTIVAL TO-DO LIST

1) Find a venue

2) Sort out security fencing

3) Sound system!!

4) Drinks/merch stand?

Stuff that needs sorting out.

First we need to pick a venue

There are three local fields where this sort of event can take place, and your choice of venue is absolutely critical. It determines how many people can come to your festival and what it will cost to make sure they all pay to get in.

Here are the three possible venues:

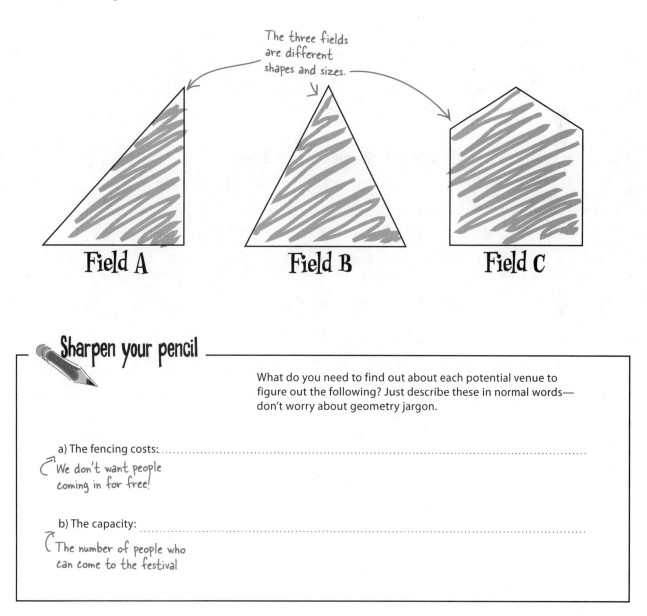

The three fields are different shapes and sizes.

Field A

Field B

Field C

Sharpen your pencil

What do you need to find out about each potential venue to figure out the following? Just describe these in normal words—don't worry about geometry jargon.

a) The fencing costs: ....

We don't want people coming in for free!

b) The capacity: ....

The number of people who can come to the festival

Sharpen your pencil
Solution

What do you need to find out about each potential venue to figure out the following? Just describe these in normal words—don't worry about geometry jargon.

a) The fencing costs: ..... The length of the edges ←———— If you wrote "perimeter," that's correct, too.

(we don't want people coming in for free!

b) The capacity ..... The area—how much space there is .....

(number of people who can come to the festival

The perimeter is the total length of the sides of a shape

If you add up the lengths of all of the sides, or edges, of a shape then you've got the perimeter. The exact calculation you need to do depends on how many sides your shape has.

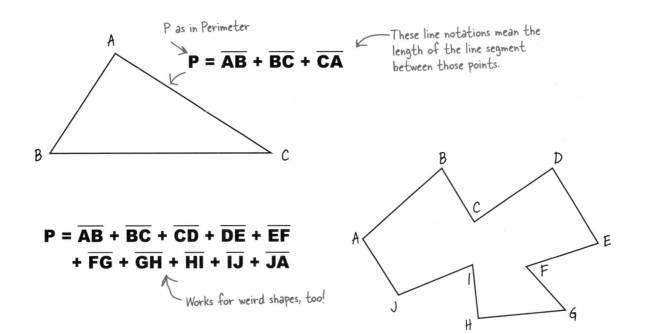

P as in Perimeter

$$P = \overline{AB} + \overline{BC} + \overline{CA}$$

These line notations mean the length of the line segment between those points.

$$P = \overline{AB} + \overline{BC} + \overline{CD} + \overline{DE} + \overline{EF}$$
$$+ \overline{FG} + \overline{GH} + \overline{HI} + \overline{IJ} + \overline{JA}$$

Works for weird shapes, too!

Fencing costs money

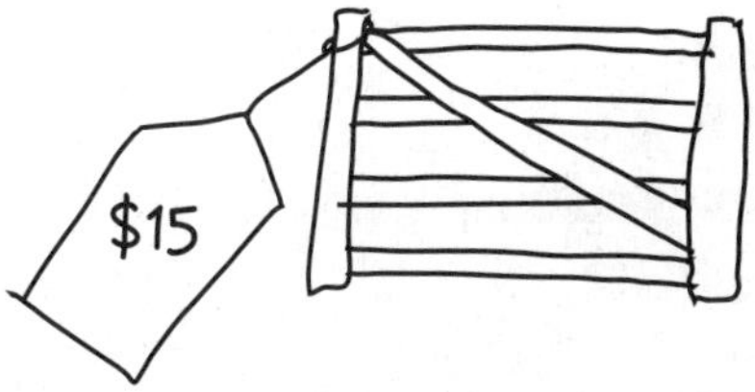

To make sure that only the people who've bought a ticket for your festival can come in, you're going to need to set up a security fence around the perimeter. And that's not cheap! The cost of setting up the security fence will be different for each potential venue.

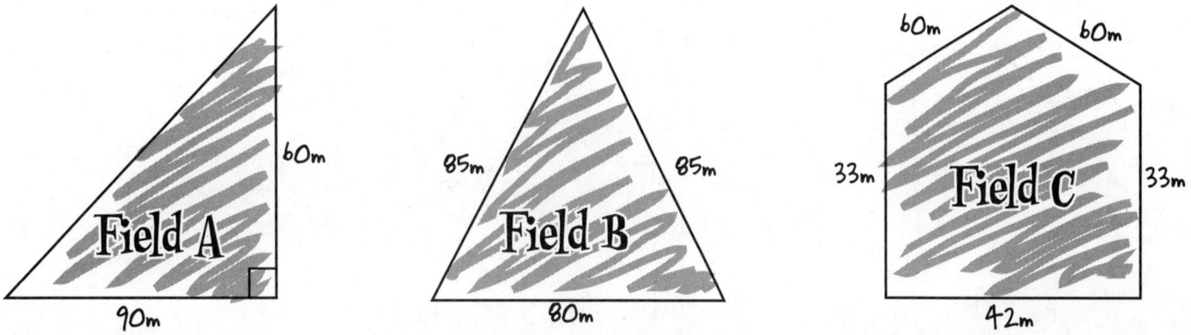

Field A
60m
90m

Field B
85m   85m
80m

Field C
60m   60m
33m   33m
42m

Exercise

Work out what it would cost to put up a security fence around each venue.
(Fencing costs $15 per meter—work by rounding UP to a whole meter.)

Use a calculator because the numbers get a bit ugly.

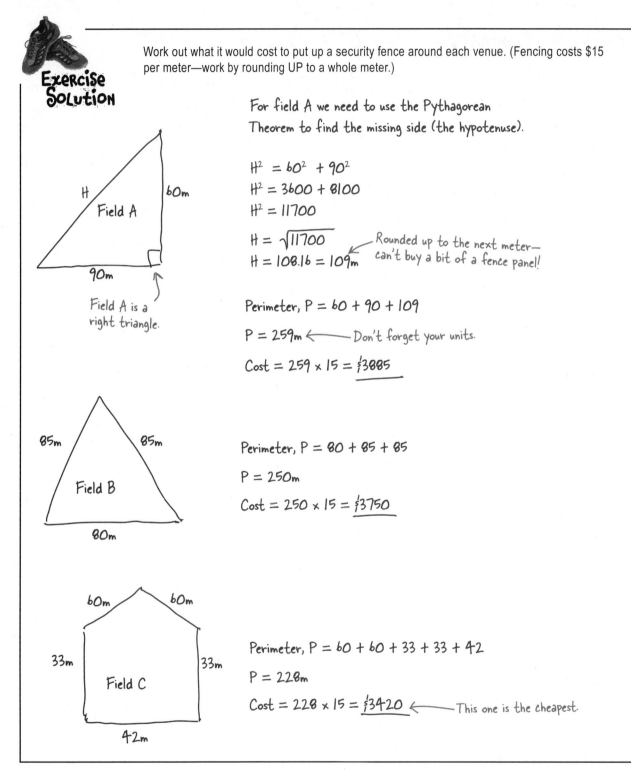

Exercise Solution

Work out what it would cost to put up a security fence around each venue. (Fencing costs $15 per meter—work by rounding UP to a whole meter.)

For field A we need to use the Pythagorean Theorem to find the missing side (the hypotenuse).

$H^2 = 60^2 + 90^2$
$H^2 = 3600 + 8100$
$H^2 = 11700$

$H = \sqrt{11700}$
$H = 108.16 = 109m$ ← Rounded up to the next meter— can't buy a bit of a fence panel!

Perimeter, $P = 60 + 90 + 109$

$P = 259m$ ← Don't forget your units.

Cost $= 259 \times 15 = \$3885$

Field A

H   60m

90m

Field A is a right triangle.

85m   85m

Field B

80m

Perimeter, $P = 80 + 85 + 85$

$P = 250m$

Cost $= 250 \times 15 = \$3750$

60m   60m

33m   33m

Field C

42m

Perimeter, $P = 60 + 60 + 33 + 33 + 42$

$P = 228m$

Cost $= 228 \times 15 = \$3420$ ← This one is the cheapest.

Does a bigger perimeter mean a bigger area?

OK, so we should go with the cheapest field to fence around—yeah?

Chris: No way! We should go with the most expensive.

Ben: Why?

Chris: Well, the one with the longest fence has to be the biggest area, and that means more ticket sales.

Tom: You think?

Chris: Makes sense to me! We should use field A for sure.

Tom: But field C is better value! It saves us over 400 bucks.

Ben: I really don't think we know which is best yet... don't you think we need to work out how many people each can hold before we decide?

BRAIN POWER

What approach would **you** take?

How many people can each venue hold?

Your buddies are sure the festival will be a sellout, and at $10 a ticket every extra person counts. Each person needs 2 square meters of space in the festival field—so how many people can each venue hold?

> Well, area of a rectangle is width times height. But none of these are rectangular...so we must need to do something different....

That's right.

The rectangle area formula isn't a bad place to start, though.

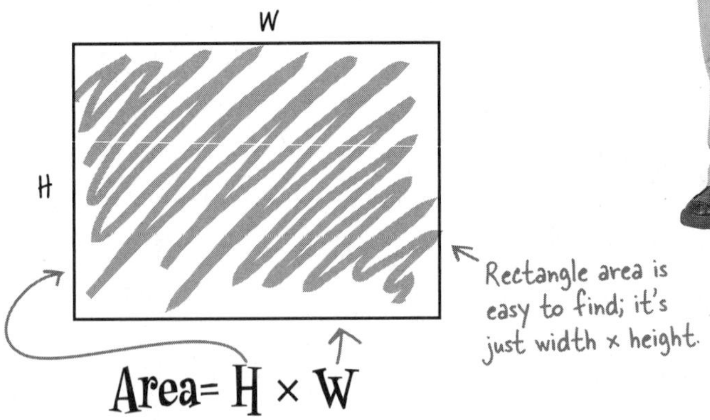

W

H

$$Area = H \times W$$

← Rectangle area is easy to find; it's just width × height.

So how does a triangle relate to a rectangle?

A triangle fits inside a bounding rectangle

A bounding rectangle is a box that fits tightly
around the triangle.

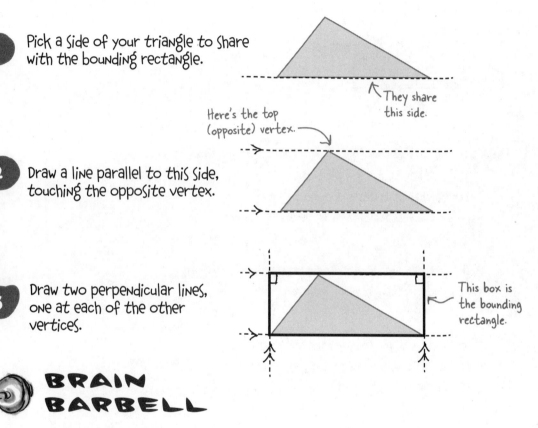

Not this sort
of bounding

To draw a bounding rectangle:

1 Pick a side of your triangle to share
with the bounding rectangle.

They share
this side.

Here's the top
(opposite) vertex.

2 Draw a line parallel to this side,
touching the opposite vertex.

3 Draw two perpendicular lines,
one at each of the other
vertices.

This box is
the bounding
rectangle.

BRAIN BARBELL

How does the shaded area of triangular field A compare with
the area of the dashed bounding rectangle surrounding it?

Scale: 1cm = 30m

60m

Field A

90m

BRAIN BARBELL

How does the shaded area of triangular field A compare with the area of the dashed bounding rectangle surrounding it?

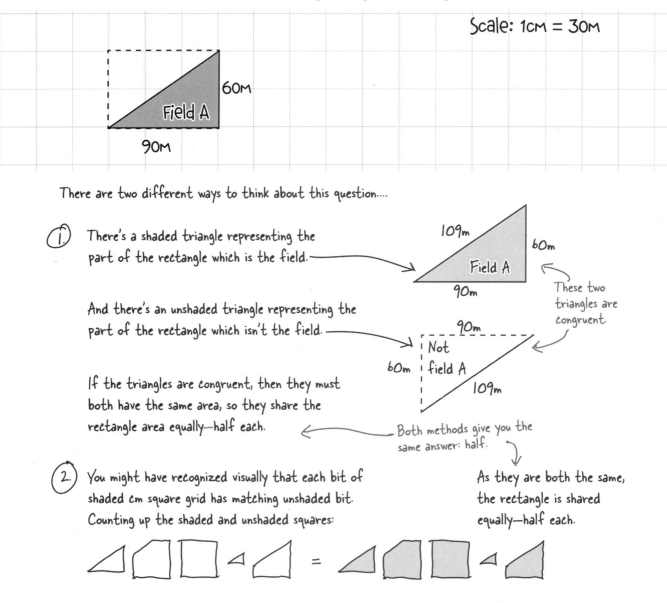

Scale: 1cm = 30m

60M

Field A

90M

There are two different ways to think about this question....

1. There's a shaded triangle representing the part of the rectangle which is the field.

109m
60m
Field A
90m

These two triangles are congruent.

And there's an unshaded triangle representing the part of the rectangle which isn't the field.

90m
Not
60m | field A
109m

If the triangles are congruent, then they must both have the same area, so they share the rectangle area equally—half each.

Both methods give you the same answer: half.

2. You might have recognized visually that each bit of shaded cm square grid has matching unshaded bit. Counting up the shaded and unshaded squares:

As they are both the same, the rectangle is shared equally—half each.

THE AREA OF THE TRIANGLE IS HALF THE AREA OF THE BOUNDING RECTANGLE.

there are no
Dumb Questions

Q: What if I've got a triangle which doesn't have a horizonal side? How do I draw a bounding rectangle?

A: Use a vertical side if you have one because it does make things easier, but if you don't have one, then any side will do.

These are both fine.

Q: We've talked about sides and angles being the same in congruent triangles, but how do you know that congruent triangles have the same area, too?

A: With a bit of flipping and rotation you can always place congruent triangles on top of one another, and they fit perfectly—so they must have the same area.

Q: What if I don't know the size of the bounding rectangle?

A: Good point. We're going to cover some ways to find that out.

For RIGHT triangles, this half-the-rectangle thing looks sensible. But what about non-right triangles?

Good question—two of our fields are not right triangles.

So what can we do?

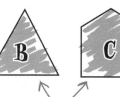

These two fields aren't right triangles.

Geometry Detective

Using a ruler, draw some non-right triangles. Use a horizontal or vertical grid line for one side (not two) of each triangle, and make the side that's on the grid line the longest side.

Draw a bounding rectangle around each triangle. How does the triangle area relate to the rectangle area?

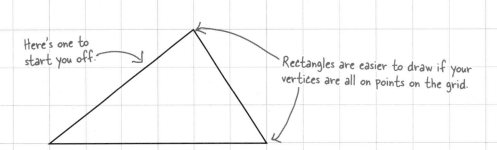

Here's one to start you off.

Rectangles are easier to draw if your vertices are all on points on the grid.

Draw at least three more
triangles on these pages

Relax

Need a 10-minute time-out?

If you're feeling kind of fuzzy about
how your triangle areas relate to rectangle areas
try having a 10-minute brain-break, then look
again before you flip the page.

Geometry Detective Solution

Using a ruler, draw some non-right triangles. Use a horizontal or vertical grid line for one side (not two) of each triangle.

Draw a bounding rectangle around each triangle. How does the triangle area relate to the rectangle area?

It's easiest to see the relationship if you split the triangle into two right triangles.

You probably didn't write this up like this—what matters is what you got as the answer.

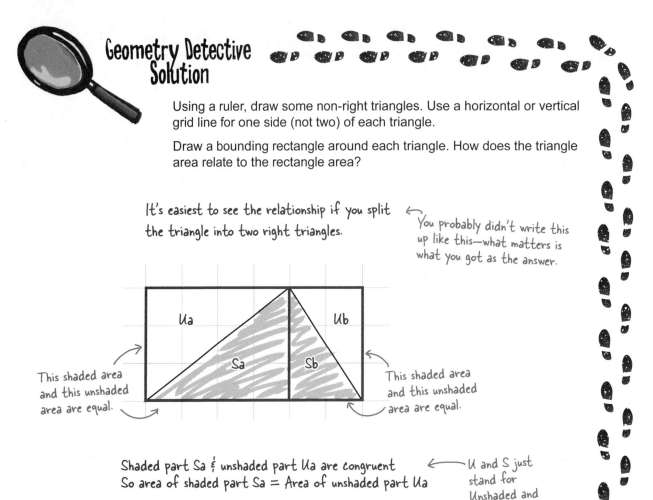

This shaded area and this unshaded area are equal.

This shaded area and this unshaded area are equal.

Shaded part Sa & unshaded part Ua are congruent
So area of shaded part Sa = Area of unshaded part Ua

U and S just stand for Unshaded and Shaded.

Shaded part Sb & unshaded part Ub are congruent
So area of shaded part Sb = Area of unshaded part Ub

Total shaded area = Sa + Sb = Ua + Ub

So—even for non-right triangles:

AREA = 1/2 THE AREA OF THE BOUNDING RECTANGLE.

This is the key thing. If you thought this—even if you didn't write it down—then you're golden.

The area of a triangle = 1/2 base x height

The area of a triangle is half the area of the bounding rectangle you could draw around it. It's usually written and found using the formula "half of base times height." The base is any one of the sides, and the height is the length of the altitude you could draw on that side to the opposite vertex (corner).

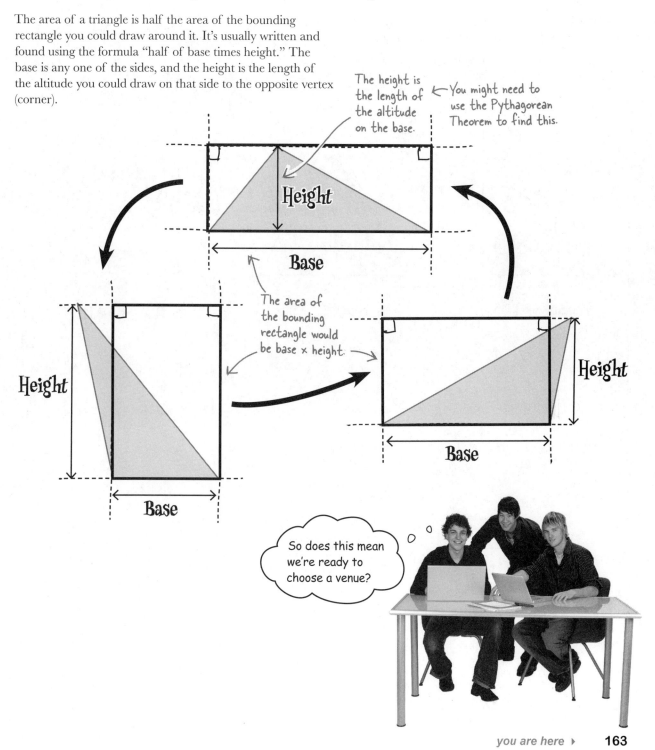

The height is the length of the altitude on the base.

← You might need to use the Pythagorean Theorem to find this.

Height

Base

The area of the bounding rectangle would be base × height.

Height

Base

Height

Base

So does this mean we're ready to choose a venue?

ℒoᴺg Exercise

Which venue gives you the most money to spend on the festival (and split the profits)?

Don't forget that you've got fencing costs. Assume it's a sellout and that each person needs 2 square meters of area.

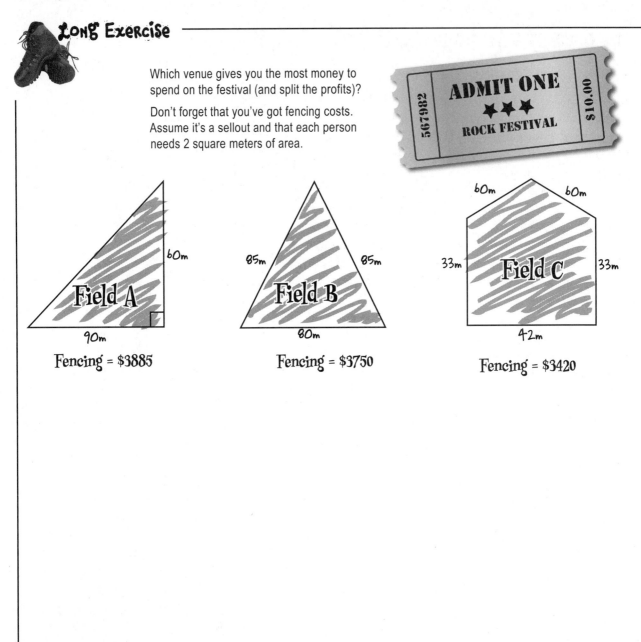

567982 ADMIT ONE ★★★ ROCK FESTIVAL $10.00

Field A
60m
90m
Fencing = $3885

Field B
85m   85m
80m
Fencing = $3750

Field C
60m   60m
33m   33m
42m
Fencing = $3420

So—which venue do you choose?

LONG EXERCISE SOLUTION

Which venue gives you the most money to spend on the festival (and split the profits!)?

Don't forget that you've got fencing costs. Assume it's a sellout and that each person needs 2 square meters of area.

ADMIT ONE
★ ★ ★
ROCK FESTIVAL

567982 $10.00

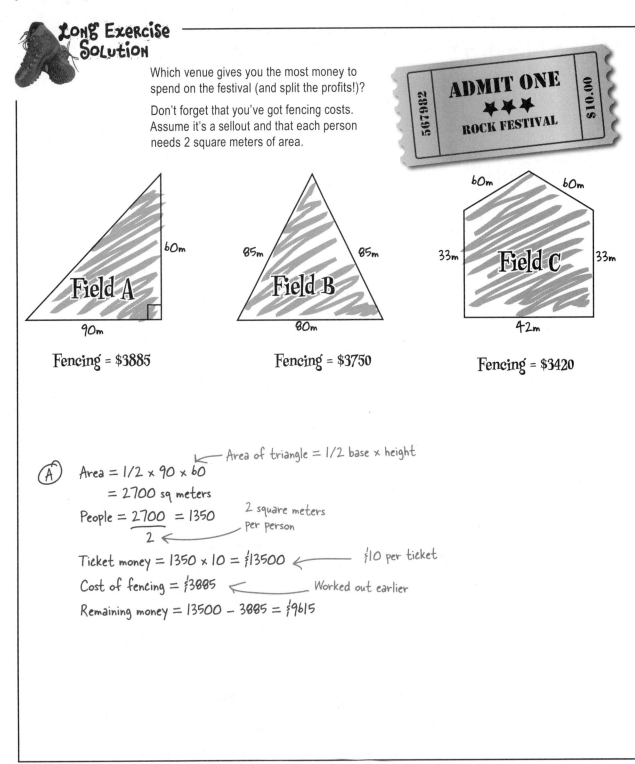

Field A

60m

90m

Fencing = $3885

Field B

85m   85m

80m

Fencing = $3750

Field C

60m   60m

33m   33m

42m

Fencing = $3420

Ⓐ   ← Area of triangle = 1/2 base × height

Area = 1/2 × 90 × 60

     = 2700 sq meters

People = $\frac{2700}{2}$ = 1350   ← 2 square meters per person

Ticket money = 1350 × 10 = $13500   ← $10 per ticket

Cost of fencing = $3885   ← Worked out earlier

Remaining money = 13500 − 3885 = $9615

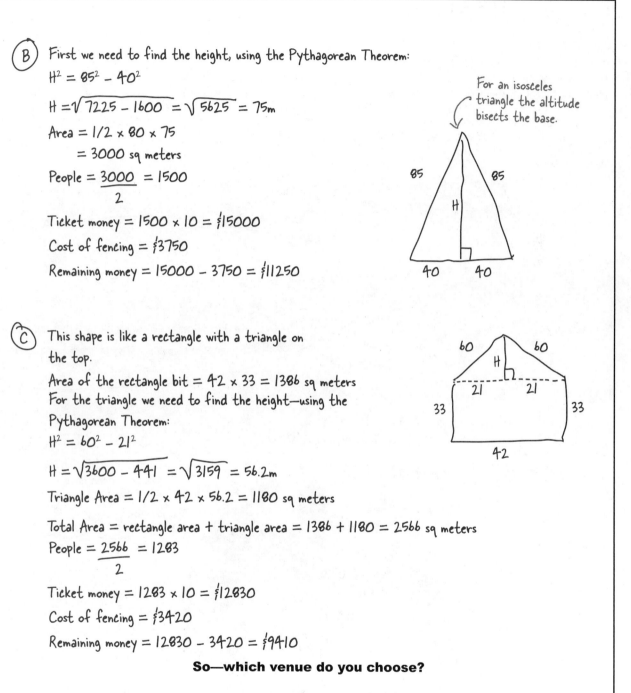

(B) First we need to find the height, using the Pythagorean Theorem:

$H^2 = 85^2 - 40^2$

$H = \sqrt{7225 - 1600} = \sqrt{5625} = 75m$

Area = 1/2 × 80 × 75

= 3000 sq meters

People = $\dfrac{3000}{2}$ = 1500

Ticket money = 1500 × 10 = $15000

Cost of fencing = $3750

Remaining money = 15000 − 3750 = $11250

For an isosceles triangle the altitude bisects the base.

(C) This shape is like a rectangle with a triangle on the top.

Area of the rectangle bit = 42 × 33 = 1386 sq meters

For the triangle we need to find the height—using the Pythagorean Theorem:

$H^2 = 60^2 - 21^2$

$H = \sqrt{3600 - 441} = \sqrt{3159} = 56.2m$

Triangle Area = 1/2 × 42 × 56.2 = 1180 sq meters

Total Area = rectangle area + triangle area = 1386 + 1180 = 2566 sq meters

People = $\dfrac{2566}{2}$ = 1283

Ticket money = 1283 × 10 = $12830

Cost of fencing = $3420

Remaining money = 12830 − 3420 = $9410

So—which venue do you choose?

Field B is the best!

You've got $11,250 to spend

After working out the capacity (number of people you can sell tickets to) and the fencing costs for each potential venue, it turns out that Field B is the best—it gives you $11,250 to spend on the festival.

Field A →
Ticket money = 1350 × 10 = $13500
Cost of fencing = $3885
Remaining money = 13500 – 3885 = $9615

This field gives you the most bucks.

Field B →
Ticket money = 1500 × 10 = $15000
Cost of fencing = $3750
Remaining money = 15000 – 3750 = $11250

Field C →
Ticket money = 1283 × 10 = $12830
Cost of fencing = $3420
Remaining money = 12830 – 3420 = $9410

Man! Over 11,000 bucks! We are gonna have the most AMAZING sound system!

TriangleGrid

Stop the press! Your buddies have secretly lined up a headline act for the festival. Solve the grid puzzle to be the first to know who the mystery band is.

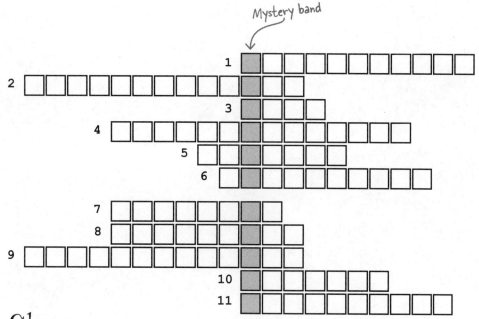

Mystery band

Clues

1. You use this theorem to find missing sides of right triangles.

2. Two angles adding up to 90° are known as this.

3. The month before August.

4. When lines cross, they create equal and opposite pairs of these.

5. If shapes have equal angles but are different sizes they are this.

6. To find common factors you can use one of these.

7. The height of a triangle is the length of this.

8. If triangles are similar and the same size then they are this.

9. Two angles adding up to 180° are know as this.

10. 3-4-5 and 5-12-13 are examples of Pythagorean ....

11. The longest side of a right triangle is called this.

Next up—picking the perfect speakers!

ROCK FESTIVAL TO-DO LIST

1) Find a venue

2) Sort out security fencing

3) Sound system!!

4) Drinks stand?

All speakers are not created equal

If you thought speakers only came in two sizes—loud and very loud—think again. The local audio store has speakers which vary by power (also called *range*)—how far the sound will travel—and by angle (also called *spread*)—how widely the sound will travel.

Sounds great!

Turn it up!

Turn it up!

Turn it up!

Sounds great!

Turn it up!

This is the spread...

...it's the same as an angle

POWER!!

Distance the sound travels from the speaker = range

...and your buddies are arguing again

Width!

Which one goes up to 11?

Power!

The guys can't decide what's really important in your speakers.

So what <u>are</u> you looking for in your speakers?

The audio store offers six different models of speakers, categorized by angle and range—how widely and how far the music will travel.

Power / ranges

Spread Angle	60m	100m	200m
60°	$1500	$2000	$3000
90°	$2500	$3000	$4000

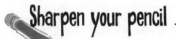

Sharpen your pencil

Think about the impact that your choice of speaker has on the festival. Fill in the table to work out what you need from your speakers. (You might not fill in every box.)

If the speakers are...	Pros (good things)	Cons (bad things)
Wider than the venue		
Narrower than the venue		
Range shorter than venue		
Range longer than venue		

Sharpen your pencil
Solution

Think about the impact that your choice of speaker has on the festival. Fill in the table to work out what you need from your speakers. (You might not fill in every box.)

If the speakers are...	Pros (good things)	Cons (bad things)
Wider than the venue	People at sides can hear	Wasted money, more expense, noise pollution ⤹ *The music would travel outside of the festival as well as inside.*
Narrower than the venue		People at edges can't hear
Range shorter than venue		People at back can't hear
Range longer than venue	People at back can hear	Wasted money, more expense, noise pollution

Don't worry if your answers are worded differently, as long as you got the general ideas down.

The ideal speakers are wider and longer than the venue...but only by a little

There's no point in paying extra so that people outside the venue can hear the show, but you need to be sure that everybody inside *can* hear.

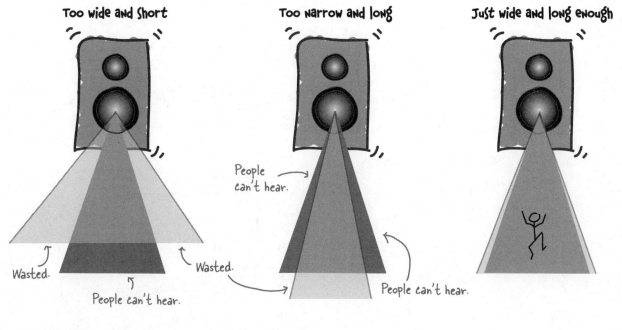

Too wide and short

Wasted.

People can't hear.

Wasted.

Too narrow and long

People can't hear.

People can't hear.

Just wide and long enough

Sharpen your pencil

Use what you know about your venue to quickly reduce your speaker options to just two. (Cross out ones you're rejecting.)

Spread Angle	60m	100m	200m	
60°	$1500	$2000	$3000	
90°	$2500	$3000	$4000	

100m will do, but can you rent the 60⁰ speaker?

The 60° speakers are way cheaper but are they going to do the job?
Will everybody in the festival be able to hear or will people at the
sides be straining to hear?

That all depends on the angle of your venue. The speakers project
sound in the shape of an isosceles triangle. Your venue is also an
isosceles triangle. If the venue angle is less than 60°, then it's all good.
Because 60° are the narrowest speakers available, it doesn't really
matter what that angle is, as long as it's less than 60°. If it's more than
60°, then you'll need to find an extra $1,000 for the 90° speakers.

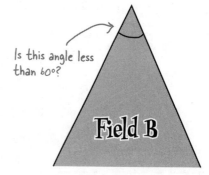

Is this angle less than 60°?

Field B

> Uh, yeah. Did
> you notice that the
> venue diagrams don't
> HAVE angles?

That's true.

The only thing you know about the venue is the side
lengths and height (you worked out the perimeter and area
earlier, too, but they don't tell you the angles either). Is that
enough to help you work out if the 60° speaker will do?

BRAIN BARBELL

Check any you think are true and that will help you find the
angle you need (it can be more than one).

☐ Sides and angles change together.

☐ From sides you can find angles accurately.

☐ From sides you can sometimes find angles accurately, and sometimes find them roughly.

☐ Sides and angles change separately.

☐ Sides tell you nothing about angles.

BRAIN BARBELL SOLUTION

Check any you think are true and that will help you find the angle you need (it can be more than one).

- ☑ Sides and angles change together.
- ☐ From sides you can find angles accurately.
- ☑ From sides you can sometimes find angles accurately, and sometimes find them roughly.
- ☐ Sides and angles change separately.
- ☐ Sides tell you nothing about angles.

Sides and height tell you a lot about angles

Imagine a triangle with three sides equal length. Do you know what the angles of that triangle would be?

The answer is upside down at the bottom of this page, but you've probably already worked it out….

Whatever...dude, can we use the 60° speakers or not?

Don't worry—he's NOT taking the stage at the festival.

Your venue, with side lengths but no angles.

85m   85m

Field B

80m

It's an equilateral triangle so the angles are each 60°

Geometry Detective

Your mission is to complete the table at the bottom of this page, by using your own hands to investigate how we can use side length and height to estimate angles.

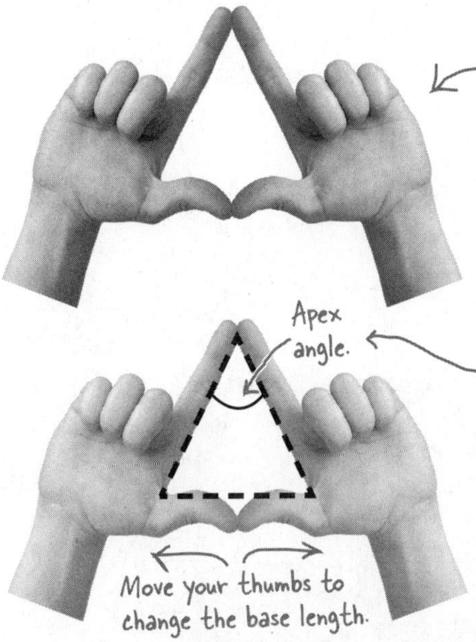

Using your forefingers and thumbs, make a triangle that is roughly equilateral.

Now change the length of the base of the triangle, by moving your thumbs apart or together. Keep your fingers the same length so that you know the triangle is isosceles.

Pay attention to how changing the proportions of the triangle—the relative length of the base and sides, and the height—changes the apex angle.

You could cut up straws or paper strips into equal lengths and try this with those, too.

Apex angle.

Move your thumbs to change the base length.

Complete the table by filling in each blank using one of the following terms:

Greater than **Less than**

Equal to **Nearly equal to**

If apex angle is	Sides and base	Height and base
Almost 0°	Side .... base	Height .... base
Less than 60°	Side .... base	Height ....1/2 base
Exactly 60°	Side .... base	Height ....1/2 base
Between 60° and 90°	Side .... base	Height ....1/2 base
Exactly 90°	Side .... base	Height ....1/2 base
More than 90°	Side .... base	Height ....1/2 base
Almost 180°	Side .... 1/2 base	Height 0

Based on your table, do you think the 60° speaker will be OK?

The 60° speakers are spot on

We know that the sides of the field are longer than the base. While this doesn't allow us to say exactly what the apex angle is—23° or 57° or something else—it does tell us that the apex angle must be **less than 60°**. That means the 60° speakers are ideal.

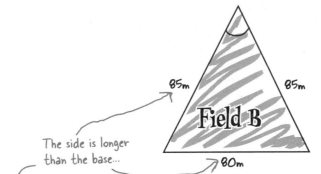

85m   85m

Field B

The side is longer
than the base...

80m

...so the triangle is in
these rows.

We already know everything we need to,
before we even compare height and base,
but we found on page 167 that the height
was 75m, so that also fits with this row.

If apex angle is	Sides and base	Height and base
Almost 0°	Side ...Greater..than... base	Height ..Greater..than. base
Less than 60°	Side ..Greater..than... base	Height .Greater..than..1/2 base
Exactly 60°	Side ...Equal..to......... base	Height ...Greater..than.1/2 base
Between 60° and 90°	Side ...Less..than....... base	Height .Greater..than..1/2 base
Exactly 90°	Side ..Less..than......... base	Height ....Equal..to......1/2 base
More than 90°	Side ...Less..than....... base	Height ...Less..than....1/2 base
Almost 180°	Side Nearly.equal..to 1/2 base	Height .Nearly.equal..to.... 0

Math geeks call this "approaching"—the
value gets real close to the thing it's nearly
equal to, but never quite gets there.

Estimating Angles Up Close

For an isoceles triangle you can use the relative values of the base, the sides, and the height to find out roughly what the apex angle (the one opposite the base) is.

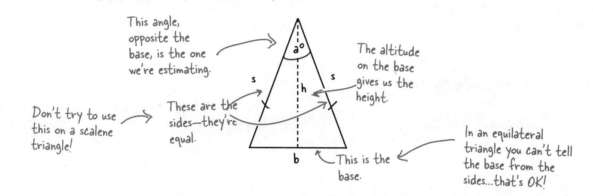

This angle, opposite the base, is the one we're estimating.

The altitude on the base gives us the height.

Don't try to use this on a scalene triangle!

These are the sides—they're equal.

In an equilateral triangle you can't tell the base from the sides...that's OK!

This is the base.

Plot sides and height against base to find which zone your apex angle is in: obtuse, acute, or very acute.

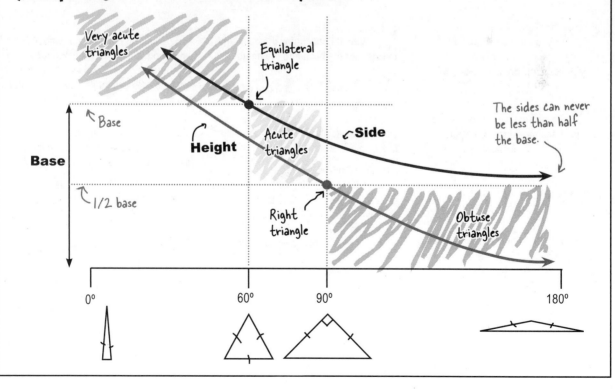

Very acute triangles

Equilateral triangle

Base

Height

Base

1/2 base

Acute triangles

Side

The sides can never be less than half the base.

Right triangle

Obtuse triangles

0°   60°   90°   180°

$2000

Budget

Tickets $15,000

Fencing −$3,750

Speakers −$2,000

_____

NET $9,250

ROCK FESTIVAL
TO-DO LIST
1) Find a venue ✓
2) Sort out security fencing ✓
3) Sound system!! ✓
4) Drinks/merch stand?

← Great choice there— →
you saved yourself
$1000!

there are no Dumb Questions

Q: Are you sure it's OK to be estimating angles like this? Don't we always need to work out exactly what the angle is?

A: If the problem or question requires a precise answer along the lines of "what is angle x," then you'll need to find the angle exactly, but sometimes it's possible to solve a problem just by knowing roughly what an angle is. For example, is it a right angle? Or maybe just is it acute or obtuse?

Q: And this is isosceles only, yeah?

A: Yes, this technique only works reliably for an isosceles triangle—where two angles and two sides are the same. An equilateral triangle is an isosceles triangle with an extra matching side, so it works for those, too, but then they're pretty easy to spot.

Q: What if I only know the base and the side, or the base and the height?

A: Sometimes that's all you need. Any time the base is greater than twice the height you know you've got an angle more than 90°. You really only need both for that tricky zone between 60° and 90°.

But don't worry—if you've got two out of three for the base, side, and height lengths, you can use the good old Pythagorean Theorem to find the other. (This only works for right and isoceles triangles—don't try it on a scalene triangle or you'll come unglued.)

Q: OK—but how do I remember this? I'm bound to get mixed up about the base and sides and and which is greater and less than which! I'm not a computer you know....

A: It's almost certainly easier to remember the zones on the graph than it is to remember the inequalities.

So, sketch that graph, mark what you know about the 60° and 90° triangles on it, and label the zones.

Q: Really? You think I could draw it just like that?

A: Go on...give it a shot on a scrap of paper now. It's a really time efficient way to check your answers in an exam by the way.

All that's left is to pick a spot for the drinks stall

The guys all agree that the best place to put the drinks
and merchandise stall is "in the center" but they're having
trouble agreeing on where the center is....

BRAIN POWER

Who's right? Where do you think the center of a triangle is?

A triangle has more than one center

Believe it or not, there are four common ways of finding the "center" of a triangle.

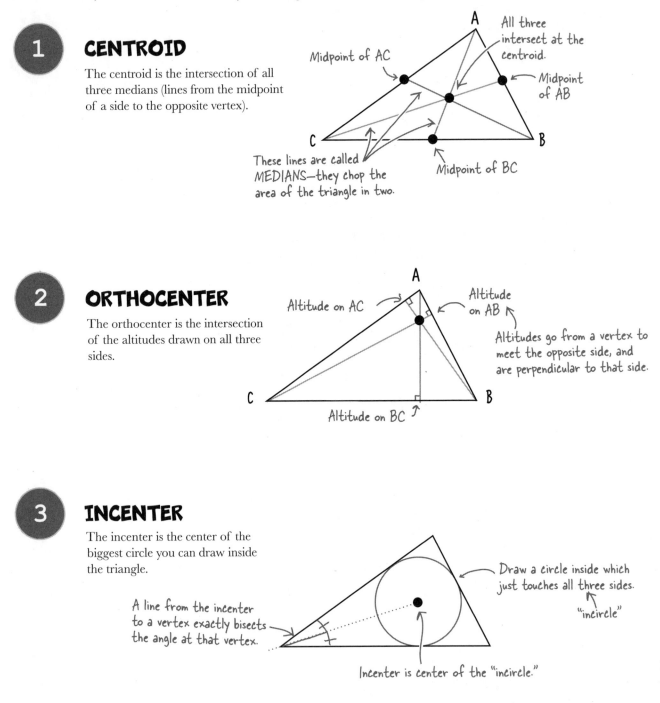

1 **CENTROID**

The centroid is the intersection of all three medians (lines from the midpoint of a side to the opposite vertex).

Midpoint of AC

All three intersect at the centroid.

Midpoint of AB

These lines are called MEDIANS—they chop the area of the triangle in two.

Midpoint of BC

2 **ORTHOCENTER**

The orthocenter is the intersection of the altitudes drawn on all three sides.

Altitude on AC

Altitude on AB

Altitudes go from a vertex to meet the opposite side, and are perpendicular to that side.

Altitude on BC

3 **INCENTER**

The incenter is the center of the biggest circle you can draw inside the triangle.

A line from the incenter to a vertex exactly bisects the angle at that vertex.

Draw a circle inside which just touches all three sides.

"incircle"

Incenter is center of the "incircle."

CIRCUMCENTER

The circumcenter is the center of the smallest circle you can draw around the outside of a triangle.

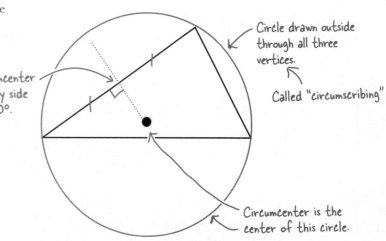

A line from the circumcenter to the midpoint of any side meets that side at 90°.

Circle drawn outside through all three vertices.

Called "circumscribing"

Circumcenter is the center of this circle.

WHO DOES WHAT?

Which kind of center was each of your buddies describing?
Pair 'em up! (You'll have one center left over.)

"The point nearest to all three fences"

"The point nearest to all three corners"

"The point where half the crowd is on either side, no matter which way you look"

Centroid

Incenter

Orthocenter

Circumcenter

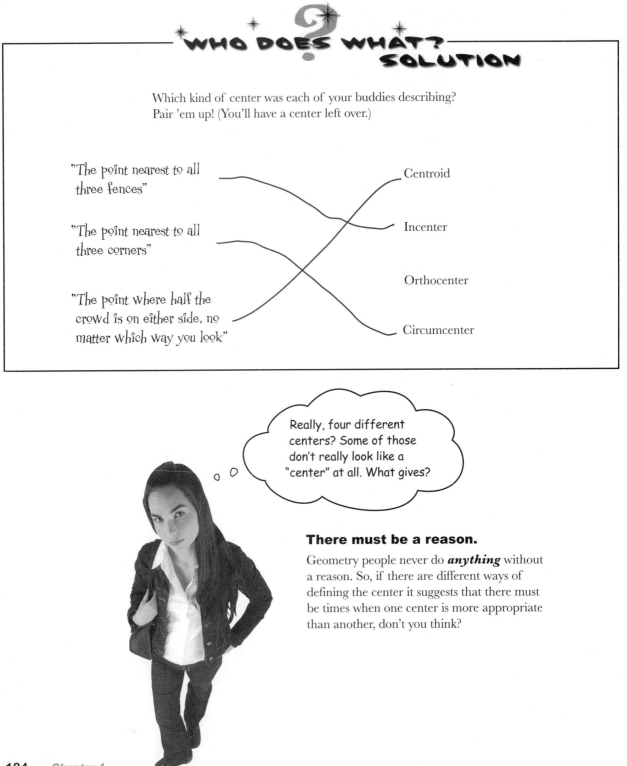

WHO DOES WHAT? SOLUTION

Which kind of center was each of your buddies describing?
Pair 'em up! (You'll have a center left over.)

"The point nearest to all three fences" — Incenter

"The point nearest to all three corners" — Circumcenter

"The point where half the crowd is on either side, no matter which way you look" — Centroid

Orthocenter

Really, four different centers? Some of those don't really look like a "center" at all. What gives?

There must be a reason.

Geometry people never do ***anything*** without a reason. So, if there are different ways of defining the center it suggests that there must be times when one center is more appropriate than another, don't you think?

Geometry Detective

Below are four different triangles. Work out (roughly) where each of the four types of "center" is for each.

Are there any definitions of "center" that wouldn't always work for placing the drinks stall?

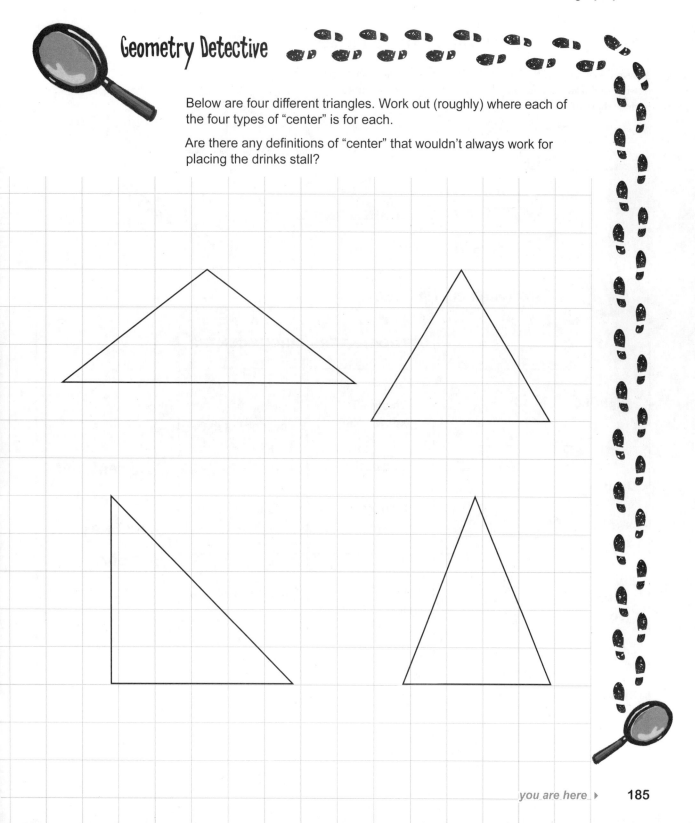

The center of a triangle can be outside the triangle

As crazy as it sounds, the circumcenter and orthocenter of a triangle can be outside the triangle. This only happens if the triangle is obtuse (one angle greater than 90°).

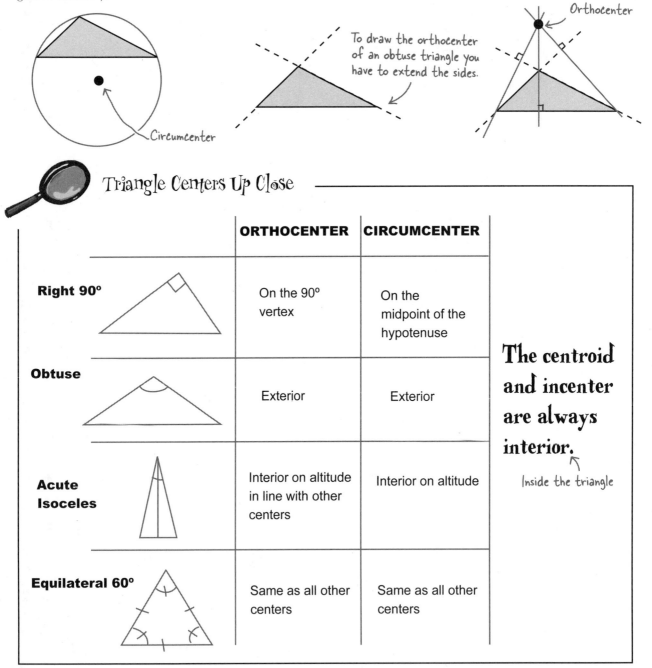

Circumcenter

To draw the orthocenter of an obtuse triangle you have to extend the sides.

Orthocenter

Triangle Centers Up Close

		ORTHOCENTER	CIRCUMCENTER
Right 90°		On the 90° vertex	On the midpoint of the hypotenuse
Obtuse		Exterior	Exterior
Acute Isoceles		Interior on altitude in line with other centers	Interior on altitude
Equilateral 60°		Same as all other centers	Same as all other centers

The centroid and incenter are always interior.

Inside the triangle

Let's put the drink stall at the centroid

Outside the venue would be a pretty weird place to put your drinks stall, so it's important to pick a center which is always interior.

The centroid also never gets pulled particularly close to one side or vertex, whereas the incenter can end up very far from one vertex if the triangle has sides of very different lengths.

An interesting thing about the centroid is that it's where the triangle balances—this means that each median splits the triangle area into two equal halves.

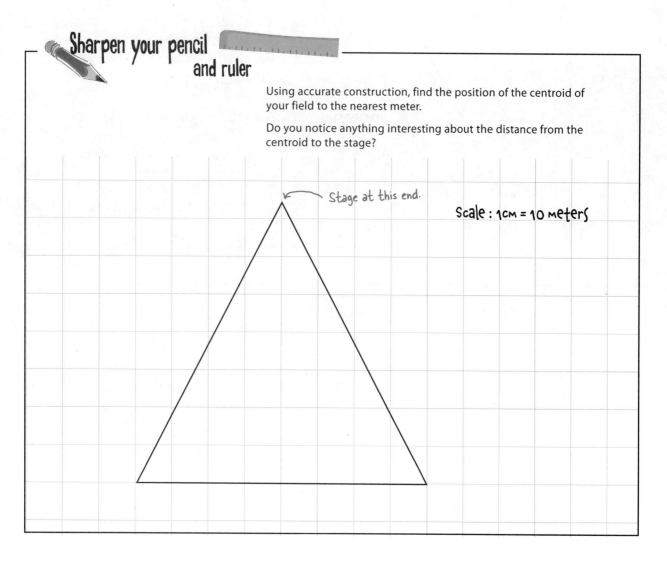

Sharpen your pencil and ruler

Using accurate construction, find the position of the centroid of your field to the nearest meter.

Do you notice anything interesting about the distance from the centroid to the stage?

Stage at this end.

Scale : 1cm = 10 meters

Sharpen your pencil
and ruler solution

Using accurate construction, find the position of the centroid of your field to the nearest meter.

Do you notice anything interesting about the distance from the centroid to the stage?

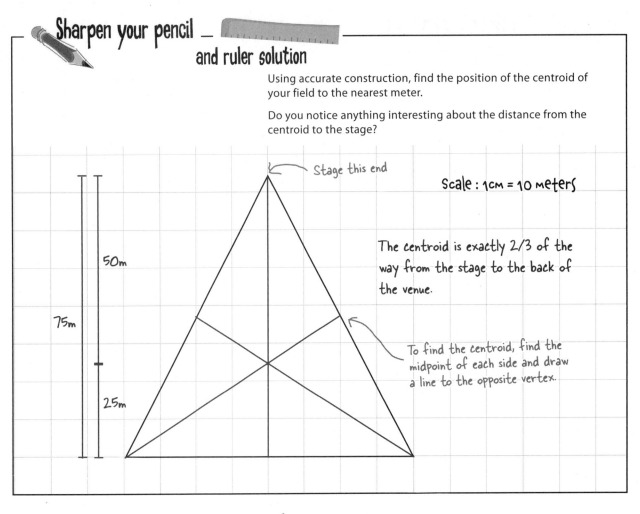

Stage this end

Scale : 1cm = 10 meters

The centroid is exactly 2/3 of the way from the stage to the back of the venue.

To find the centroid, find the midpoint of each side and draw a line to the opposite vertex.

50m

75m

25m

there are no Dumb Questions

Q: I drew an equilateral triangle and my centers were all in the same place—did I do it wrong?

A: That's perfect! All the four centers of an equilateral triangle are in the same place. It's the only type of triangle that truly has a "middle." Centers of an isoceles triangle all fall on the altitude of the base—the line you might use to find the height.

Q: Do I really need to know four different centers—won't one do?

A: There are actually over 3,500! These are the four you'll use the most, though. The centroid is used to balance a triangle in the physical world, but the other three all have uses. Each has different potential to find other things—a line from the incenter to any vertex of a triangle bisects that angle at that vertex, so you can use the incenter to find that angle if you need to.

Q: Why on earth would a center that lies outside the triangle be useful?

A: The orthocenter and circumcenter only lie outside the triangle **sometimes**. If your triangle is acute then you might have need for a center that is exactly the same distance from all three corners, or one where a line to any side meets it at a right angle. It's totally dependent on what triangle you've got and what you're trying to do with it.

The rock festival is ready!

1 You chose a venue based on area/capacity minus the perimeter fencing costs.

2 You picked the right speakers to rent—saving yourself $1,400!

3 You found the centroid of the field—the perfect spot for the drinks and merchandise.

Dude, we sold ALL the tickets! Are you ready to ROCK?

An extra one of your buddies is onboard to help out with the logistics.

The people behind the drinks stall won't see the stage...

The engineers have started setting up, and now that the drink stand is in place, your buddies realize that it blocks the view for whoever will be standing behind it.

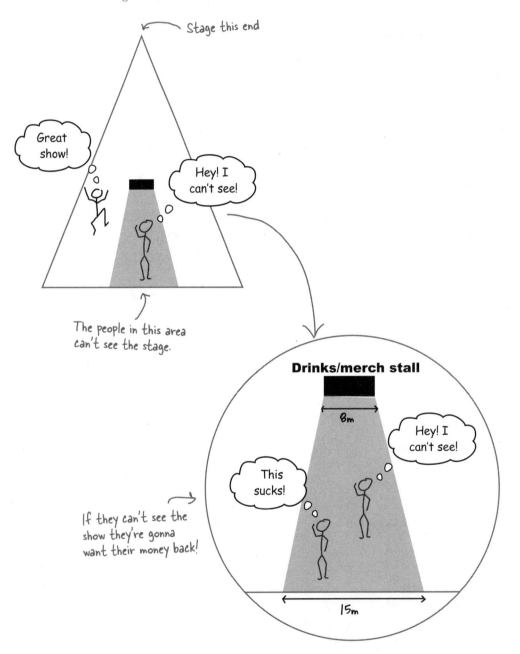

Oh, no! We're gonna have to give refunds to people who get stuck in that area. Or rope it off and sell fewer tickets....

Ben →

Chris ↓

Dave →

← **Tom**

Tom: Ugh. How many refunds?

Dave: I don't know…we'd need to work out how many people are in that area behind the screen.

Ben: Can't we do something else? People are gonna be so disappointed!

Chris: And it means less cash as well—which means we'll have less to split between us all!

Tom: I guess we could hire a giant screen maybe? But I bet they are way expenisve.

Dave: But we'll lose money if we give refunds anyway—so anything less than that amount is worth it, yes?

Sharpen your pencil

If you had to give refunds to all the people in that area, how much would it cost you? (1 person per 2 square meters and $10 a ticket.)

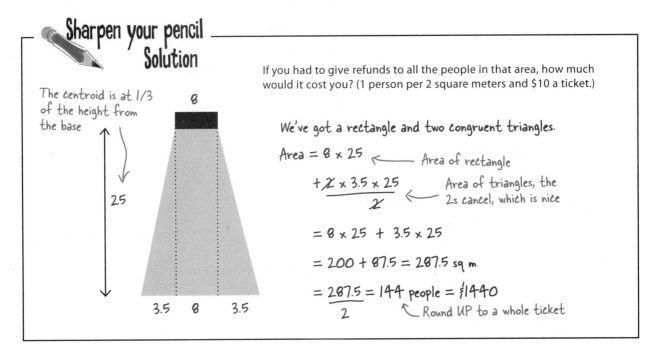

Sharpen your pencil
Solution

The centroid is at 1/3 of the height from the base

8

25

3.5   8   3.5

If you had to give refunds to all the people in that area, how much would it cost you? (1 person per 2 square meters and $10 a ticket.)

We've got a rectangle and two congruent triangles.

Area = 8 × 25 ←——— Area of rectangle

$+ \dfrac{\cancel{2} \times 3.5 \times 25}{\cancel{2}}$ ←—— Area of triangles, the 2s cancel, which is nice

= 8 × 25 + 3.5 × 25

= 200 + 87.5 = 287.5 sq m

$= \dfrac{287.5}{2} = 144$ people = $1440

↖ Round UP to a whole ticket

You need a screen for less than $1,440

The cost of giving refunds to people behind the drinks stall would be $1,440. If you can find a screen to rent for less than that, it would let everybody see the show and you'd still have some profits to split.

Ouch—screens start at like $2,000! Wait…this video store has a special offer on some slightly damaged screens—that could save us!

Sharpen your pencil
Solution

Use what you know about your venue to quickly reduce your speaker options to just two. (Cross out ones you're rejecting.)

The venue length is 75m.

Spread Angle	60m	100m	200m
🔊 60°	~~$1500~~	$2000	~~$3000~~
🔊 90°	~~$2500~~	$3000	~~$4000~~

Both these speakers are going to reach the back.

---

there are no
Dumb Questions

Q: **Is this for real? Are speakers really sold by angles?**

A: Yup. Common sizes are 60°, 90°, and 120°. Bass speakers tend to be broad, and tweeters (for higher frequencies) are narrower, but all vary. Apparently for perfect sound quality you need to sit so that you and your two stereo speakers (left and right) make a perfect equilateral triangle. But that's only for really hardcore music geeks.

Q: **And don't we need more than one speaker?**

A: What you're really ordering is a speaker system, which would probably have more than one unit.

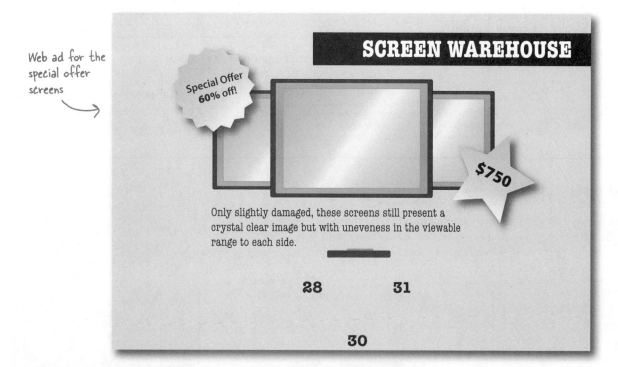

Web ad for the special offer screens →

SCREEN WAREHOUSE

Special Offer 60% off!

$750

Only slightly damaged, these screens still present a crystal clear image but with unevenness in the viewable range to each side.

28   31

30

Will the special offer screen still do the job?

The screen is suitable if the viewable range (r) reaches all the way to the back—but you only have the sides given on the specification in the ad.

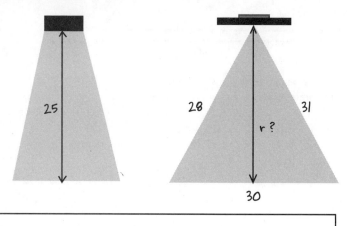

25

28   31

r ?

30

Sharpen your pencil

What technique from your Geometry Toolbox can help you find the screen's viewable distance, r?

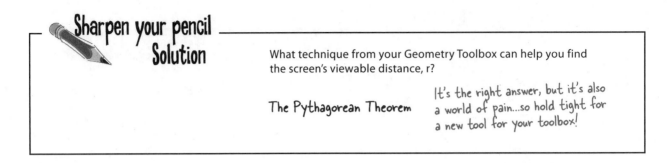

Sharpen your pencil
Solution

What technique from your Geometry Toolbox can help you find the screen's viewable distance, r?

The Pythagorean Theorem

It's the right answer, but it's also a world of pain...so hold tight for a new tool for your toolbox!

The screen viewing area is a <u>scalene triangle</u>

This means that when you add an altitude, it doesn't bisect the base. Instead of creating two nice, neat **congruent** right triangles, you get two **different** right triangles. And you don't know the base of either of them!

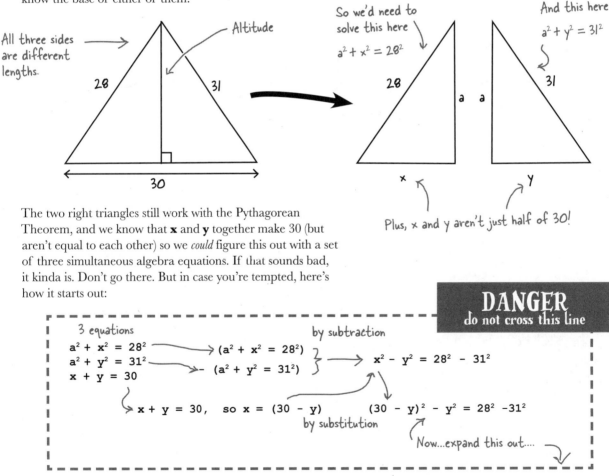

All three sides are different lengths.

Altitude

28

31

30

So we'd need to solve this here

$a^2 + x^2 = 28^2$

And this here

$a^2 + y^2 = 31^2$

28

a   a

31

x

y

Plus, x and y aren't just half of 30!

The two right triangles still work with the Pythagorean Theorem, and we know that **x** and **y** together make 30 (but aren't equal to each other) so we *could* figure this out with a set of three simultaneous algebra equations. If that sounds bad, it kinda is. Don't go there. But in case you're tempted, here's how it starts out:

DANGER
do not cross this line

3 equations

$a^2 + x^2 = 28^2$

$a^2 + y^2 = 31^2$

$x + y = 30$

by subtraction

$(a^2 + x^2 = 28^2)$

$- (a^2 + y^2 = 31^2)$

$x^2 - y^2 = 28^2 - 31^2$

$x + y = 30$,   so $x = (30 - y)$

by substitution

$(30 - y)^2 - y^2 = 28^2 - 31^2$

Now...expand this out.....

Wouldn't it be dreamy if there was a way to find the height of a scalene triangle without simultaneous equations? But I know it's just a fantasy....

You can find area from sides using <u>Hero's formula</u>

Fortunately there is a formula known as Hero's formula, or sometimes Heron's formula, which lets you find the area of a triangle when you only know the sides. First you find the semi-perimeter—half the perimeter—and then you just pop the numbers into your equation and wham, there's your area. Phew!

Aptly named, saving you from three variable simultaneous equations!

Hero's formula:

$$\text{Area} = \sqrt{s\,(s-a)\,(s-b)\,(s-c)}$$

$$s = \frac{a+b+c}{2}$$

Semiperimeter

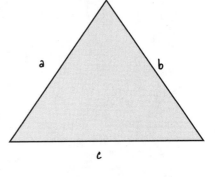

1   ## Find the semiperimeter

The semiperimeter is exactly what it sounds like.

Just add the three side lengths together to get the perimeter, and then divide it by 2.

$$s = \frac{a+b+c}{2}$$

Do this bit first...

...then pop your answer in here, replacing "s."

2   ## Use the main formula

The main formula has four *s*'s in it. You use the value you got for the semiperimeter for each *s*. Don't forget to get the square root of the whole thing when you're done.

$$\text{Area} = \sqrt{s\,(s-a)\,(s-b)\,(s-c)}$$

Fascinating. And really useful I'm sure, if you're trying to find **area**. But weren't we trying to find the **height** of that triangle? I don't think a detour into weird area formulas is what we need.

Actually, it could be *exactly* what we need!

One of the neat things about geometry is the way ***everything links up***. Like back in Chapter 1, when you proved Benny was innocent because you could find the angle two different ways, and they didn't match up.

Hero's formula doesn't exist in a vacuum—it fits into all the tools you already have in your Geometry Toolbox. And that's why it's a **big** help in finding not just area, but height as well.…

BRAIN BARBELL

How could Hero's formula for triangle area also make it easy for you to find the triangle's height?

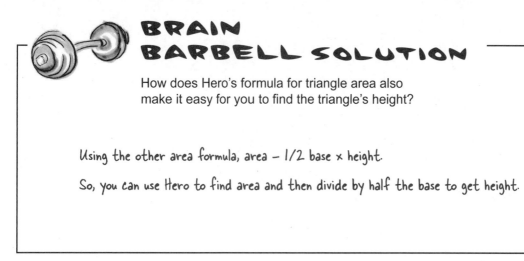

BRAIN BARBELL SOLUTION

How does Hero's formula for triangle area also make it easy for you to find the triangle's height?

Using the other area formula, area – 1/2 base x height.

So, you can use Hero to find area and then divide by half the base to get height.

Hero's formula and "1/2 base x height" work together

If you know three sides of a scalene triangle, you can use Hero's formula to find the area, and then use the formula you already know to find the height.

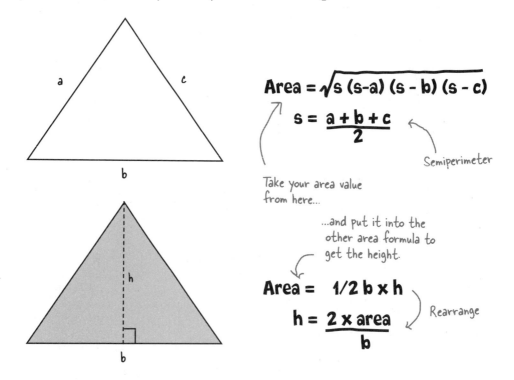

$$\text{Area} = \sqrt{s\,(s-a)\,(s-b)\,(s-c)}$$

$$s = \frac{a+b+c}{2}$$

Semiperimeter

Take your area value from here...

...and put it into the other area formula to get the height.

$$\text{Area} = \tfrac{1}{2}\,b \times h$$

$$h = \frac{2 \times \text{area}}{b}$$

Rearrange

there are no Dumb Questions

Q: So do I actually ever need to do that gnarly three simultaneous equations thing?

A: No. It would work though, so if you ever can't remember Hero's formula, it could get you out of a jam! But yeah—forget it. Sorry we did that to you....

Q: Can I use Hero's formula for isoceles and right triangles, too?

A: For a right triangle, 1/2 base x height is always easier as the two sides on the right angle give you base and height. For an isoceles triangle it's up to you which you find easier.

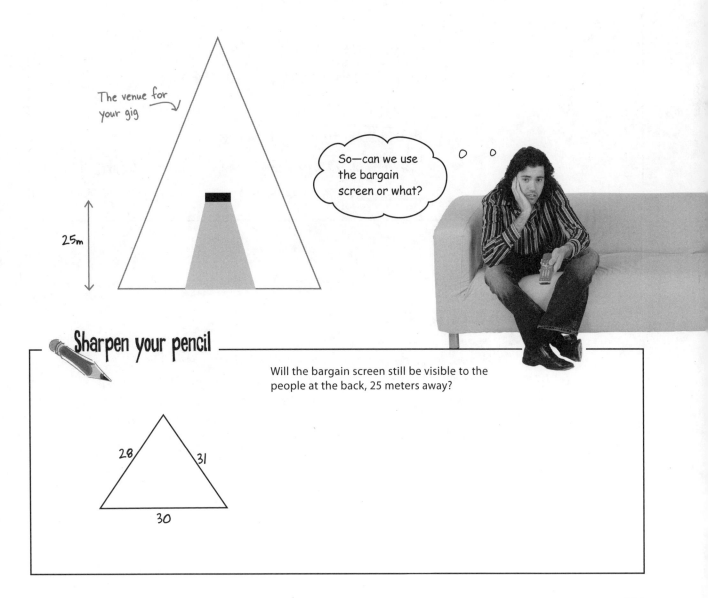

The venue for your gig

25m

So—can we use the bargain screen or what?

Sharpen your pencil

Will the bargain screen still be visible to the people at the back, 25 meters away?

28   31

30

Sharpen your pencil
Solution

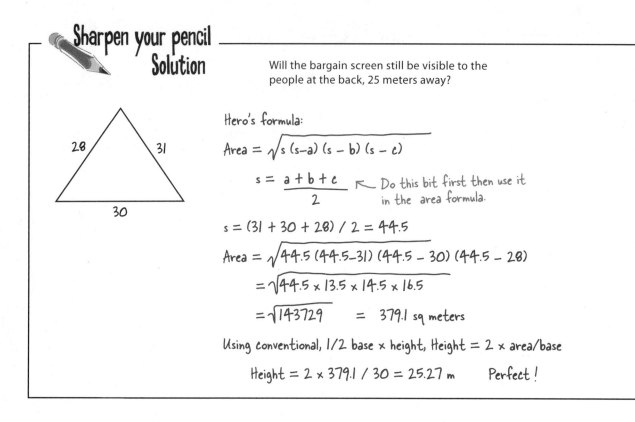

Will the bargain screen still be visible to the people at the back, 25 meters away?

Hero's formula:

$$Area = \sqrt{s\,(s-a)\,(s-b)\,(s-c)}$$

$$s = \frac{a + b + c}{2}$$ ← Do this bit first then use it in the area formula.

$$s = (31 + 30 + 28) / 2 = 44.5$$

$$Area = \sqrt{44.5\,(44.5-31)\,(44.5 - 30)\,(44.5 - 28)}$$

$$= \sqrt{44.5 \times 13.5 \times 14.5 \times 16.5}$$

$$= \sqrt{143729} \quad = \quad 379.1 \text{ sq meters}$$

Using conventional, 1/2 base × height, Height = 2 × area/base

$$Height = 2 \times 379.1 / 30 = 25.27 \text{ m} \qquad Perfect!$$

BULLET POINTS

- Combine tools from your toolbox to get the answer you need.

- There's sometimes more than one way to solve a problem—pick the way that seems like the least amount of work!

- Use the relationship between sides and height to estimate angles.

- Draw sketches or graphs, or use your hands if you're stuck remembering what those side-height-base relationships are.

The rock festival is gonna...rock!

Great venue? Check!

Fencing? Check!

Awesome sound? Check!

Giant video screen? Check!

Sold out? Check!

Way to go! Now—go hang out backstage with the headline act—**Pajama Death**. They're legendary!

Budget

Tickets $15,000

Fencing −$3,750

Speakers −$2,000

Screen −$750

_____

NET $8,500

Still plenty left for lights, fireworks, backstage parties—excellent!

Awesome!!!

Pajama Death are headlining. Wicked!

ADMIT ONE
★★★
ROCK FESTIVAL
567982

ADMIT ONE
★★★
ROCK FESTIVAL
$10.00

Your Geometry Toolbox

You've got Chapter 4 under your belt and now you've added properties of triangles to your toolbox. For a complete list of tool tips in the book, head to www.headfirstlabs.com/geometry.

Area = 1/2 base × height

The height comes from the altitude of the triangle—it's perpendicular to the base.

Hero's formula

$$Area = \sqrt{s\,(s-a)\,(s-b)\,(s-c)}$$

$$s = \frac{a+b+c}{2}$$

Find the semiperimeter first, then use it to find area.

Perimeter

Add up all the sides of your shape to find the perimeter.

Circumcenter

The center of a circle drawn outside the triangle, through every vertex

Centroid

The point where all three medians of the triangle intersect

Orthocenter

The point where all three altitudes of the triangle intersect

Incenter

The center of a circle drawn inside the triangle, touching every side

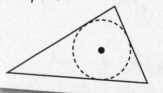

Median

A line from the midpoint of a side to the opposite vertex

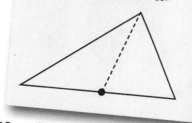

TriangleGrid Solution

Stop the press! Your buddies have secretly lined up a headline act for the festival. Solve the grid puzzle to be the first to know who the mystery band are.

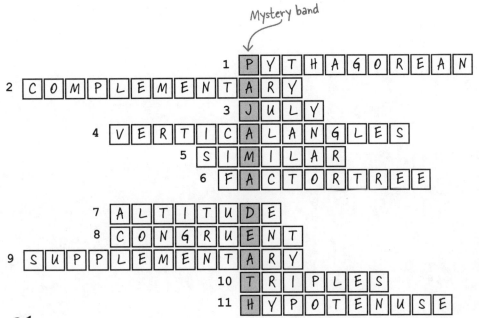

Mystery band ↓

1. P Y T H A G O R E A N
2. C O M P L E M E N T A R Y
3. J U L Y
4. V E R T I C A L A N G L E S
5. S I M I L A R
6. F A C T O R T R E E
7. A L T I T U D E
8. C O N G R U E N T
9. S U P P L E M E N T A R Y
10. T R I P L E S
11. H Y P O T E N U S E

Clues

1. You use this theorem to find missing sides of right triangles.

2. Two angles adding up to 90° are known as this.

3. The month before August.

4. When lines cross, they create equal and opposite pairs of these.

5. If shapes have equal angles but are different sizes they are this.

6. To find common factors you can use one of these.

7. The height of a triangle is the length of this.

8. If triangles are similar and the same size then they are this.

9. Two angles adding up to 180° are know as this.

10. 3-4-5 and 5-12-13 are examples of Pythagorean ....

11. The longest side of a right triangle is called this.

Are you a Pajama Death fan? Help the band get their own TV show in Head First Algebra.

5 circles

Going round and round

And if I push you real hard, this time you'll go over the top and travel the whole circumference. Now, hold tight!

OK, life doesn't have to be so straight after all!

There's no need to reinvent the wheel, but aren't you glad you're able to use it? From cars to rollercoasters, many of the **most important solutions** to life's problems rely on *circles* to get the job done. Free yourself from straight edges and pointy corners—there's no end to the curvy possibilities once you master **circumference**, **arcs**, and **sectors**.

It's not just pizza—it's war!

Everybody knows it—***you want pizza? Go to Mario's!*** But times are changing. The MegaSlice chain is muscling in on Mario's turf, and to make matters worse it looks like they're fighting dirty!

Mario is upset.

My pizzeria is in trouble! The MegaSlice chain is stealing my customers, but their special offer claims are lies!

The MegaSlice promo flyer

25% MORE PIZZA

SPECIAL OFFER!

MEGASLICE $10 DEAL

10 Megaslices for the same price as 8 Mario-Slices

How does MegaSlice's deal measure up?

MegaSlice claims you get more pizza with their 10-slices-for-$10 deal.
You certainly get more slices, but does that mean more pizza?

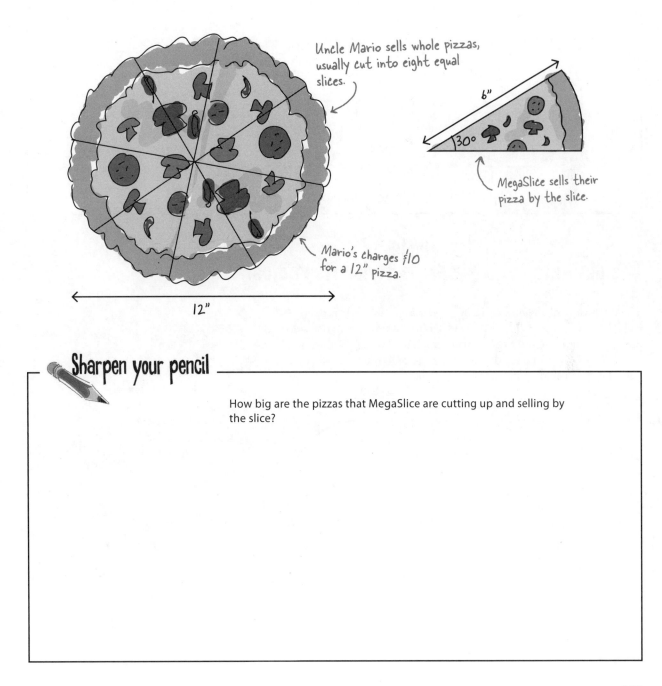

Uncle Mario sells whole pizzas, usually cut into eight equal slices.

6"

30°

MegaSlice sells their pizza by the slice.

Mario's charges $10 for a 12" pizza.

12"

Sharpen your pencil

How big are the pizzas that MegaSlice are cutting up and selling by the slice?

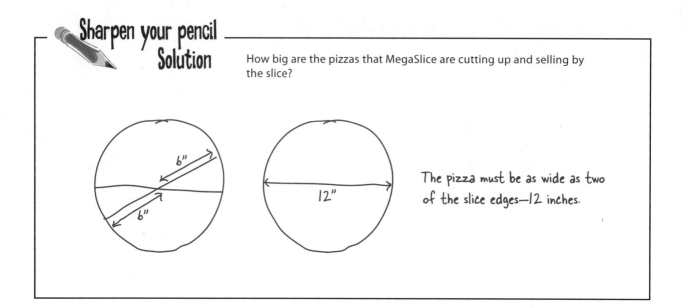

Sharpen your pencil
Solution

How big are the pizzas that MegaSlice are cutting up and selling by the slice?

The pizza must be as wide as two of the slice edges—12 inches.

The diameter of a ~~pizza~~ circle, is twice its radius

The total width of the pizza—12 inches—is a special circle property we call **diameter**. The diameter of a circle is the distance from one side to the other, passing through the center. The length of the edge of the pizza slice is called **radius**, and is the distance from the center to any point on the edge. Diameter is always two times the radius.

It doesn't matter which points on the edge you draw your diameter between, or where you draw the radius to, it's always the same size (just remember to include the center).

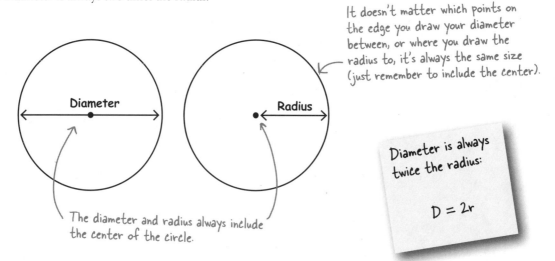

The diameter and radius always include the center of the circle.

Diameter is always twice the radius:

$$D = 2r$$

How do slices compare to whole pizzas?

MegaSlice's deal is in slices, and Mario's is for a whole pie. But MegaSlice's 10 slices also come from a 12-inch pizza. So—does that mean that their deal is better, because it's 10 slices and not just 8?

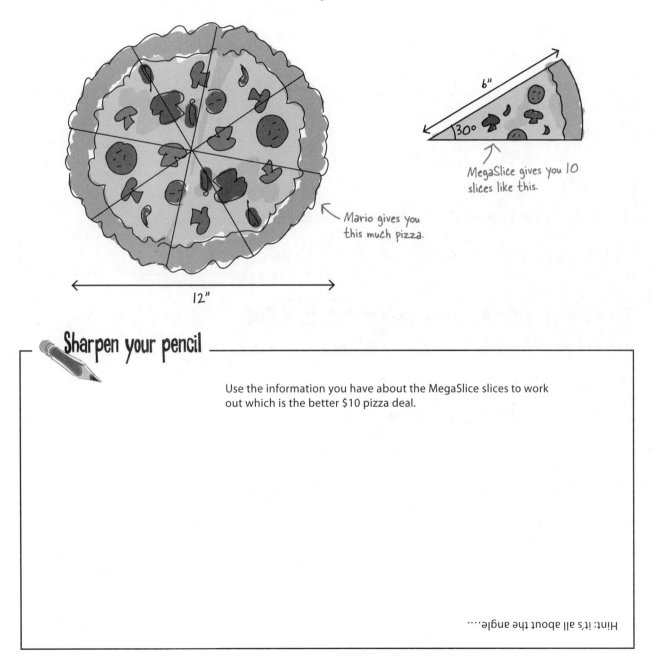

6"

30°

MegaSlice gives you 10 slices like this.

Mario gives you this much pizza.

12"

Sharpen your pencil

Use the information you have about the MegaSlice slices to work out which is the better $10 pizza deal.

Hint: it's all about the angle....

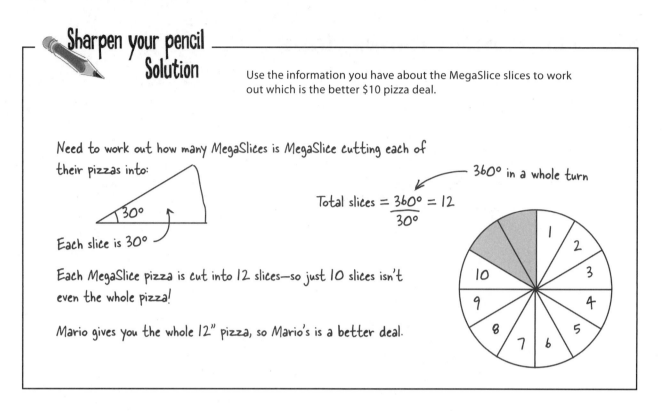

Sharpen your pencil Solution

Use the information you have about the MegaSlice slices to work out which is the better $10 pizza deal.

Need to work out how many MegaSlices is MegaSlice cutting each of their pizzas into:

30°

Each slice is 30°

360° in a whole turn

Total slices = $\frac{360°}{30°}$ = 12

Each MegaSlice pizza is cut into 12 slices—so just 10 slices isn't even the whole pizza!

Mario gives you the whole 12" pizza, so Mario's is a better deal.

Sectors of a circle have angles totaling 360°

When a circle is divided up using lines that pass through the center of the circle, the resulting "pizza slice" shapes are called ***sectors***. The MegaSlice sectors each have 30° angles, so a full pizza from them would have to be 12 slices (30° × 12 = 360°). But they only give you 10 slices, which means their deal gives you less than a whole 12" pizza!

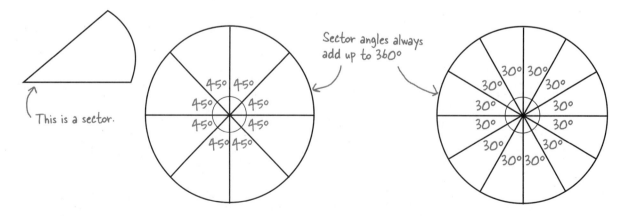

This is a sector.

Sector angles always add up to 360°

MegaSlice's $10 deal is a con!

MegaSlice is only offering 10 of the 30° slices in a whole 12" pizza, so while they're offering a greater number of slices than Uncle Mario, it's actually less pizza overall than his deal! Now you've got the hard facts, there's no way MegaSlice can keep claiming to be a better value.

From: MegaSlice CEO
To: You
cc: MegaSlice Marketing, MegaSlice Legal

Thank you for bringing the honest mistake in our $10 deal promotional materials to our attention. We have withdrawn the posters and leaflets concerned.

Sincerely, MegaSlice CEO

MegaSlice is eating humble pie.

Thanks! MegaSlice stopped their lies and business is recovering—slowly. But I have an idea to rev things up! You like pepperoni, right?

Pepperoni crust pizza—but at what price?

Mario's idea is to offer a pepperoni crust on any pizza. He wants to offer the best possible value to his customers so he'll only add the minimum price needed to cover the cost of the extra pepperoni. Sounds simple, right? But Mario has a problem with the sequence of events in the pizza process:

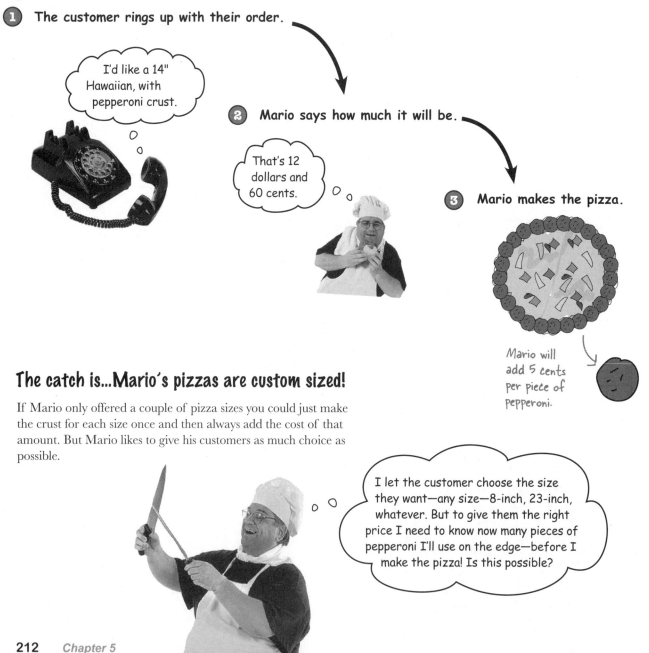

1 **The customer rings up with their order.**

I'd like a 14" Hawaiian, with pepperoni crust.

2 **Mario says how much it will be.**

That's 12 dollars and 60 cents.

3 **Mario makes the pizza.**

Mario will add 5 cents per piece of pepperoni.

The catch is...Mario's pizzas are custom sized!

If Mario only offered a couple of pizza sizes you could just make the crust for each size once and then always add the cost of that amount. But Mario likes to give his customers as much choice as possible.

I let the customer choose the size they want—any size—8-inch, 23-inch, whatever. But to give them the right price I need to know now many pieces of pepperoni I'll use on the edge—before I make the pizza! Is this possible?

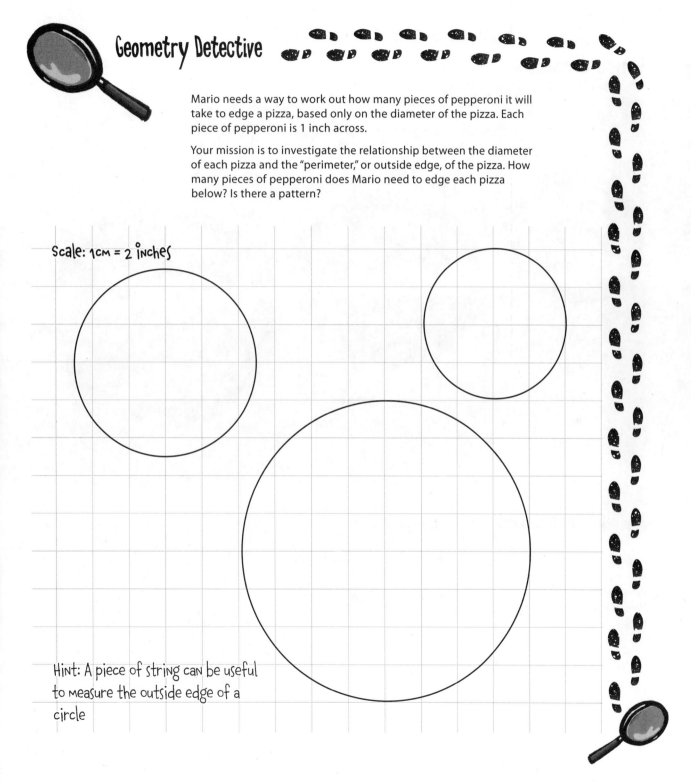

Geometry Detective

Mario needs a way to work out how many pieces of pepperoni it will take to edge a pizza, based only on the diameter of the pizza. Each piece of pepperoni is 1 inch across.

Your mission is to investigate the relationship between the diameter of each pizza and the "perimeter," or outside edge, of the pizza. How many pieces of pepperoni does Mario need to edge each pizza below? Is there a pattern?

Scale: 1cm = 2 inches

Hint: A piece of string can be useful to measure the outside edge of a circle

The pepperoni perimeter is 3 (and a bit) times diameter

There's a pattern: no matter what size the pizza, the pepperoni perimeter is a little more than three times the diameter.

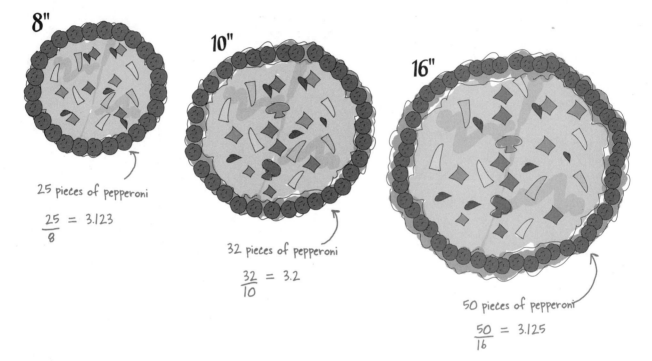

8"

25 pieces of pepperoni

$$\frac{25}{8} = 3.123$$

10"

32 pieces of pepperoni

$$\frac{32}{10} = 3.2$$

16"

50 pieces of pepperoni

$$\frac{50}{16} = 3.125$$

A circle's circumference is diameter x Pi

In geometry terms, a circle's perimeter is called the ***circumference***, and no matter how big or small that circle is, you can find the circumference by multiplying the diameter by the same number. Roughly it's just barely over three, and we call this number **Pi**. Pi is usually written using this symbol:

π

Pi

The circumference of a circle is given by:

Circumference = πD

or

Circumference = $2\pi r$

These are equal because $D = 2r$.

Ha ha...very funny. Not. The chapter is about pizza and now you're throwing in a made-up number called "Pie"....

"Pi" is actually a letter from the Greek alphabet.

It's used to stand in for the number that you always get when you divide a circle's circumference by its diameter.

You can probably find a Pi button on your calculator. It might give you 3.14159, or it might give you something with even more decimal places.

BRAIN POWER

Why do you think "math people" find Pi, or π, more useful than 3.14159265358979323846264433832795...?

No calculator? Substitute the fraction 22/7 for a close approximation to Pi.

there are no
Dumb Questions

Q: When you showed the formula for circumference, you put πD but also 2πr—does it matter which I use?

A: D is diameter, and r is radius, so the two formulas always give the same answer since D = 2r. It doesn't matter which you use—it just depends on whether you've got the diameter or the radius available.

Q: So, in like two pages you've shown Pi as "just barely over three," 3.14159, a way-long decimal, and even 22/7... which is it?

A: Pi is what is known as an *irrational* number. That means that even if you used a billion decimal places you couldn't write it down completely. Depending on what you're using Pi for you might need to use a more or less precise version of it. On an exam, you'll usually be told what to use, otherwise just take whatever your calculator gives you when you press the π button, and remember to round your answer to fit with the question.

OK, so Pi is really nothing more than a placeholder for the ratio between a circle's circumference and its diameter? I guess it's quicker than writing or saying 3.14159265358979 32384626433832795...?

Exactly.

Math geeks talk about Pi as if it's kind of magic—and it's certainly pretty useful because it works for every circle, every time—but really Pi is just a quick way of saying "that big long number that you get when you divide circumference by diameter."

Mario wants to put your pepperoni crust pricing formula to the test

So, is this "Pi" thing just theoretical, or does it actually work in the real world? It's time to put your formula to the test.

I can't be charging my customers wrong prices—understand? So, we'll have a go and see if it works. I tell you the orders, you work out the pepperoni edge prices, and then when I make the pizza we'll see if you were right!

Exercise

Mario wants to charge 5 cents per piece of pepperoni. How much does he need to add to the price of the following orders if they all want the new pepperoni crust? (Use your calculator and round up or down to the nearest whole piece of pepperoni.)

1: A 14" TexMex with extra jalapenos

2: A 20" "The Works" (hold the mushrooms)

3: Two 6" kids' Hawaiians, one with extra pineapple

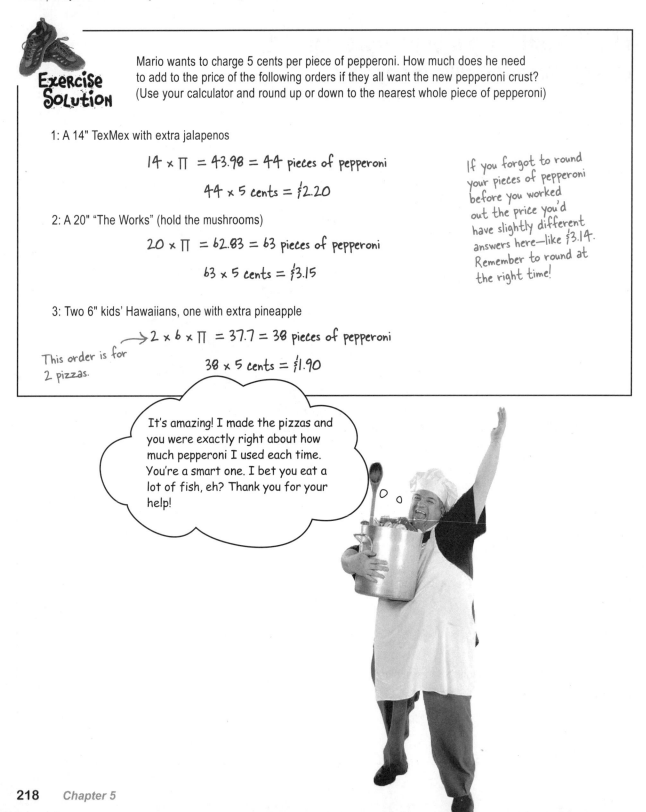

Exercise
Solution

Mario wants to charge 5 cents per piece of pepperoni. How much does he need
to add to the price of the following orders if they all want the new pepperoni crust?
(Use your calculator and round up or down to the nearest whole piece of pepperoni)

1: A 14" TexMex with extra jalapenos

$$14 \times \pi = 43.98 = 44 \text{ pieces of pepperoni}$$

$$44 \times 5 \text{ cents} = \$2.20$$

2: A 20" "The Works" (hold the mushrooms)

$$20 \times \pi = 62.83 = 63 \text{ pieces of pepperoni}$$

$$63 \times 5 \text{ cents} = \$3.15$$

If you forgot to round
your pieces of pepperoni
before you worked
out the price you'd
have slightly different
answers here—like $3.14.
Remember to round at
the right time!

3: Two 6" kids' Hawaiians, one with extra pineapple

$$\rightarrow 2 \times 6 \times \pi = 37.7 = 38 \text{ pieces of pepperoni}$$

This order is for
2 pizzas.

$$38 \times 5 \text{ cents} = \$1.90$$

It's amazing! I made the pizzas and
you were exactly right about how
much pepperoni I used each time.
You're a smart one. I bet you eat a
lot of fish, eh? Thank you for your
help!

The customers are always ~~right~~ fussy

Pizza is for sharing, right? But some of Mario's customers want to share a pizza and only have pepperoni crust on PART of it.

I want at least a little pepperoni with my pizza!!

I'm vegetarian, and pepperoni upsets Fido's stomach.

Nobody likes a pukey dog....

Gina definitely doesn't want pepperoni.

Fido doesn't really care what kind of crust he gets, but Gina says no pepperoni for him, either.

Todd & Gina agree on everything except pepperoni crust.

Sharpen your pencil

Todd and Gina always order an 18-inch Super Veggie deep-dish pizza. They ask for it cut into nine equal slices. Gina eats three, Todd has the other six, but he always gives his last half-slice to Fido.

How much extra should Mario charge for their complicated pepperoni crust requirements?

Sharpen your pencil
Solution

Todd and Gina always order an 18-inch Super Veggie deep-dish pizza. They ask for it cut into nine equal slices. Gina eats three, Todd has the other six, but he always gives his last half-slice to Fido.

How much extra should Mario charge for their complicated pepperoni crust requirements?

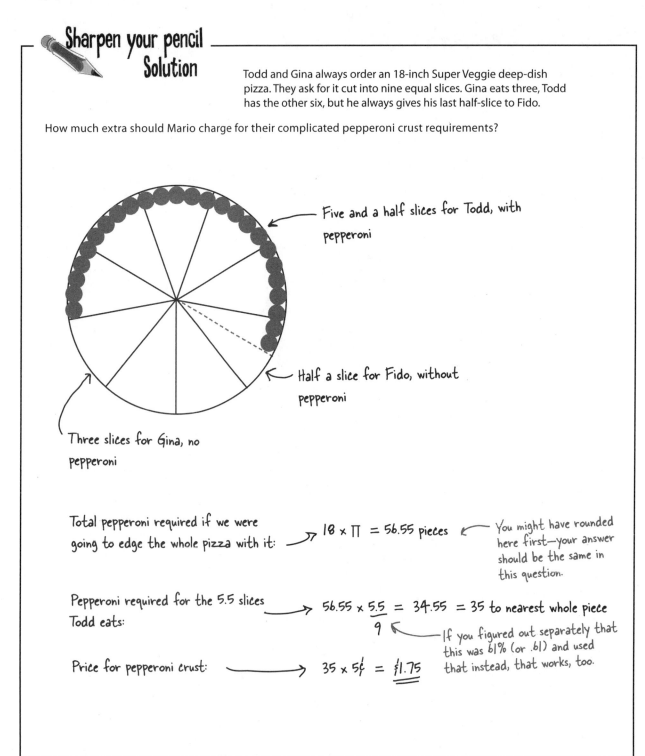

Five and a half slices for Todd, with pepperoni

Half a slice for Fido, without pepperoni

Three slices for Gina, no pepperoni

Total pepperoni required if we were going to edge the whole pizza with it: $\longrightarrow$ $18 \times \Pi = 56.55$ pieces $\longleftarrow$ You might have rounded here first—your answer should be the same in this question.

Pepperoni required for the 5.5 slices Todd eats: $\longrightarrow$ $56.55 \times \dfrac{5.5}{9} = 34.55 = 35$ to nearest whole piece $\longleftarrow$ If you figured out separately that this was 61% (or .61) and used that instead, that works, too.

Price for pepperoni crust: $\longrightarrow$ $35 \times 5¢ = \underline{\$1.75}$

An arc is a section of the circumference

Covering part of the pizza crust with pepperoni creates an **arc**. That's geometry jargon for a section of the circumference of a circle—whether it goes almost all the way around a circle, or just a tiny part of it.

Arcs can be described by the proportion of the circumference they cover—a semicircle covers half—or by the angle of the corresponding sector (the arc is the outside curved part of the sector).

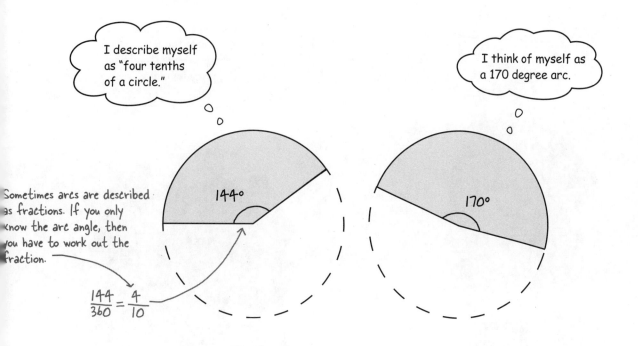

> I describe myself as "four tenths of a circle."

> I think of myself as a 170 degree arc.

Sometimes arcs are described as fractions. If you only know the arc angle, then you have to work out the fraction.

$$\frac{144}{360} = \frac{4}{10}$$

144°

170°

Arc length is circumference x sector angle / 360°

To find the length of an arc, first you need to find the circumference of the circle the arc is part of. That means that you need either the diameter or radius of the circle. Then you can use the sector angle to find out the length of the part of the circumference that your arc covers.

This calculation helps you find the fraction of the circle that the sector represents. If you already know it's 2/5 or 1/12, then just use the fraction instead!

$$\textbf{Arc Length} = \textbf{2}\boldsymbol{\pi}\textbf{r} \times \frac{\textbf{Sector Angle}}{\textbf{360°}}$$

Mario's business is booming!

Mario's customers love being able to choose how much of their pizza
has pepperoni crust. Even when they can't agree on pepperoni, they
can agree that Mario's is the perfect pizza to suit them.

> Mario's just so
> flexible—and a great
> value, too....

> We ordered a 30-inch pizza
> to share, and only the pitchers
> wanted pepperoni crust, and he
> gave us the price instantly!

BULLET POINTS

- Circumference $= \pi D = 2\pi r$

- Diameter $= 2r$

- Arc Length $= 2\pi r \times \dfrac{\text{Sector Angle}}{360°}$

But MegaSlice is at it again...

Just when Mario thought MegaSlice had learned its lesson, MegaSlice is back with a TV commercial claiming their restaurants give you more for your money. And on the surface it sounds like a pretty good deal.

Two 12-inch MegaSlice pizzas for the same price as ONE 18-inch Mario's pizza! A way better value for your family!

Help me! I know it's not true, but sales are still falling. If we can't fight this I'll close for sure....

Sharpen your pencil

What property of the pizzas do you need to compare to be able to work out whether MegaSlice's claims are true?

Sharpen your pencil Solution

What property of the pizzas do you need to compare to be able to work out whether MegaSlice's claims are true?

Area

If you put "Volume" here, you're right, but as all Mario's pizzas have the same thickness, area will also cover it. More on that in Chapter 7!

We need to find the area of the two pizza deals

MegaSlice offers two 12" pizzas for the same price as one of Mario's 18" pizzas.

So, which deal is more AREA of pizza, because that's the property you actually get to eat.

OK, you could kind of nibble the circumference if you wanted....

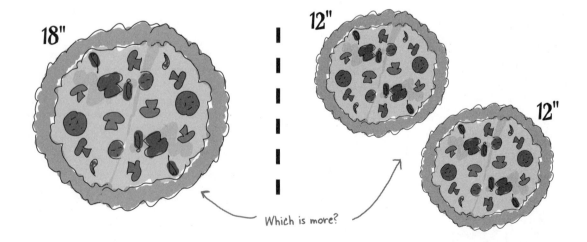

18"   12"   12"

Which is more?

Frank   Jim   Joe

Well, I'm not sure we even need to work this one out. Two times twelve is more than eighteen, so MegaSlice's deal is more pizza.

Frank: Not so fast. I'm not sure it's that simple.

Joe: What's hard about it? 12 + 12 = 24. 24 is way more than 18. Game over.

Jim: But that's not area—that's diameter!

Joe: Yeah—and?

Jim: Well, I'm not sure that you can just add up diameters and then compare them instead of comparing area.

Joe: Why not? Anyway, we don't even know how to find the area of a circle. Hey—is this making you guys hungry?

Frank: Shut up! This is serious. We need to find a way to compare the actual pizza area, and not just the diameters.

Jim: Agreed, but where do we start?

Frank: Well, I guess we should just start from what we do know and go from there.

BRAIN BARBELL

Which technique from your Geometry Toolbox could help you work out the pizza areas approximately?

Triangle area = 1/2 base × height

Similarity and congruence

Angles in a quadrilateral add up to 360°.

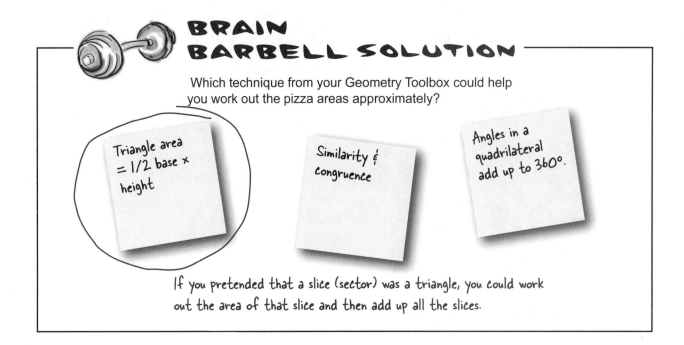

BRAIN
BARBELL SOLUTION

Which technique from your Geometry Toolbox could help
you work out the pizza areas approximately?

Triangle area
= 1/2 base ×
height

Similarity &
congruence

Angles in a
quadrilateral
add up to 360°.

If you pretended that a slice (sector) was a triangle, you could work
out the area of that slice and then add up all the slices.

Each sector (slice) is a triangle (kind of)

A sector of a circle isn't actually a triangle with three straight edges,
but it's pretty close. You can find the approximate area of that sector
by using the radius of the circle as the triangle height, and the arc
length as the triangle base.

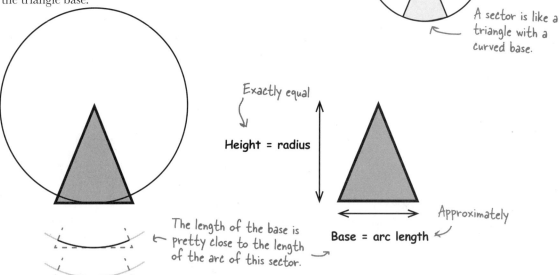

A sector is like a
triangle with a
curved base.

Exactly equal

Height = radius

The length of the base is
pretty close to the length
of the arc of this sector.

Base = arc length

Approximately

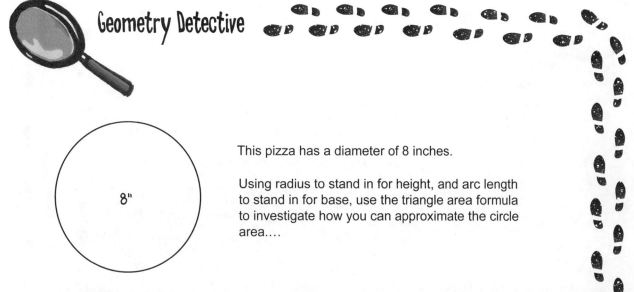

Geometry Detective

This pizza has a diameter of 8 inches.

Using radius to stand in for height, and arc length to stand in for base, use the triangle area formula to investigate how you can approximate the circle area....

8"

1. By chopping the pizza base into 6 equal slices

2. By chopping the pizza base into 10 equal slices

3. By chopping the pizza base into 30 equal slices

What do you notice as you chop it into more and more slices?

Geometry Detective Solution

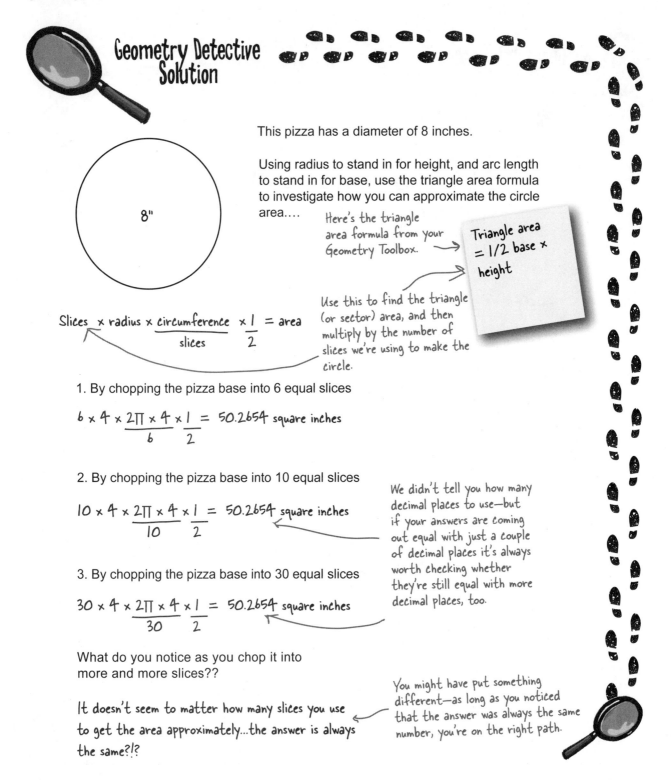

This pizza has a diameter of 8 inches.

Using radius to stand in for height, and arc length to stand in for base, use the triangle area formula to investigate how you can approximate the circle area....

8"

Here's the triangle area formula from your Geometry Toolbox.

Triangle area = 1/2 base × height

Use this to find the triangle (or sector) area, and then multiply by the number of slices we're using to make the circle.

$$\text{Slices} \times \text{radius} \times \frac{\text{circumference}}{\text{slices}} \times \frac{1}{2} = \text{area}$$

1. By chopping the pizza base into 6 equal slices

$$6 \times 4 \times \frac{2\pi \times 4}{6} \times \frac{1}{2} = 50.2654 \text{ square inches}$$

2. By chopping the pizza base into 10 equal slices

$$10 \times 4 \times \frac{2\pi \times 4}{10} \times \frac{1}{2} = 50.2654 \text{ square inches}$$

We didn't tell you how many decimal places to use—but if your answers are coming out equal with just a couple of decimal places it's always worth checking whether they're still equal with more decimal places, too.

3. By chopping the pizza base into 30 equal slices

$$30 \times 4 \times \frac{2\pi \times 4}{30} \times \frac{1}{2} = 50.2654 \text{ square inches}$$

What do you notice as you chop it into more and more slices??

It doesn't seem to matter how many slices you use to get the area approximately...the answer is always the same?!?

You might have put something different—as long as you noticed that the answer was always the same number, you're on the right path.

No way! That's just freaky. How come they all came out the same?

Frank: Well…I think I know HOW it happened, I just don't know WHY it happened.…

Joe: Really? Say more.…

Frank: Hmm. Well, see how the first time we multiplied by six because we were chopping it into six slices, and then we divided by six while calculating the arc length? They cancel out, don't they?

Jim: Yeah—and I guess the next one does as well—when we chopped it into 10, and we divided and multiplied by 10.

Frank: And with the 30 pieces, too. The number of slices is always cancelling out.

Joe: OK, I see that, but don't you think it's kind of weird that we got the same answer each time, even though we were using arc length just as an approximation for the triangle base?

Frank: Ah! OK—that's it—that's exactly it!

Joe: What's it?

Frank: OK, when we have a few big slices the arc length is really pretty different from the triangle base.

Jim: Yes, very different.

Frank: But, the more slices we chop the circle into, the nearer the arc length is to the triangle base length. So, eventually—if we chop the circle into like a million pieces—it's not an approximation anymore, it's a perfect fit!

Jim: Yes. Cool! In fact, I think we've just found the formula for the area of a circle.

BRAIN BARBELL

Based on the pattern you found in your investigation, what do YOU think the formula for the area of a circle might be?

Try starting with a word equation, and then work it up in algebra if you're feeling confident.

BRAIN
BARBELL SOLUTION

Based on the pattern you used in your investigation, what do YOU think the formula for the area of a circle might be?

The pattern for finding area based on triangle area was:

$$\text{area} = \text{slices} \times \text{radius} \times \frac{\text{circumference}}{\text{slices}} \times \frac{1}{2}$$

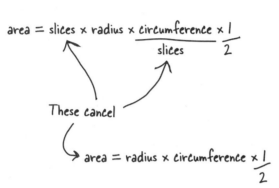

These cancel

$$\text{area} = \text{radius} \times \text{circumference} \times \frac{1}{2}$$

You might have stopped here, which is great work, or you might have gone a step further....

$$\text{Circumference} = \Pi D = 2\Pi r$$

$$\text{area} = \text{radius} \times \text{circumference} \times \frac{1}{2}$$

$$\text{area} = \text{radius} \times \cancel{2}\Pi r \times \frac{1}{\cancel{2}}$$

$$\text{area} = r \times \Pi r = \Pi r^2$$

Area of a circle = πr²

It's true—if you crammed enough narrow triangles into a circle, eventually you'd get a perfect fit. You'd need an infinite number of triangles, but it turns out the number of triangles cancels neatly, so wham—you just found the formula for the area of a circle!

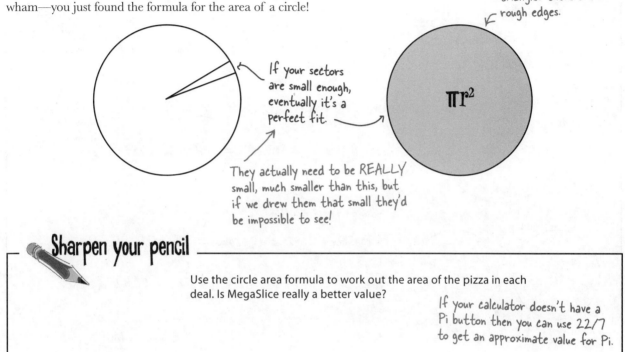

With enough (infinite) triangles there are no rough edges.

If your sectors are small enough, eventually it's a perfect fit.

They actually need to be REALLY small, much smaller than this, but if we drew them that small they'd be impossible to see!

Sharpen your pencil

Use the circle area formula to work out the area of the pizza in each deal. Is MegaSlice really a better value?

If your calculator doesn't have a Pi button then you can use 22/7 to get an approximate value for Pi.

Sharpen your pencil
Solution

Use the circle area formula to work out the area of the pizza in each deal. Is MegaSlice really a better value?

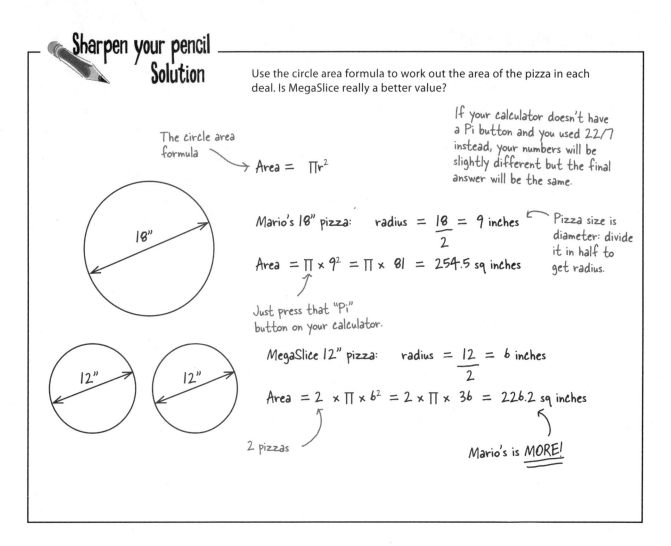

If your calculator doesn't have a Pi button and you used 22/7 instead, your numbers will be slightly different but the final answer will be the same.

The circle area formula

→ Area = πr^2

Mario's 18" pizza: radius = $\frac{18}{2}$ = 9 inches

Pizza size is diameter: divide it in half to get radius.

Area = $\pi \times 9^2 = \pi \times 81 = 254.5$ sq inches

Just press that "Pi" button on your calculator.

MegaSlice 12" pizza: radius = $\frac{12}{2}$ = 6 inches

Area = 2 × $\pi \times 6^2 = 2 \times \pi \times 36 = 226.2$ sq inches

2 pizzas

Mario's is <u>MORE!</u>

---

there are no Dumb Questions

Q: Eek—$2\pi r$ and πr^2 look pretty similar don't you think? Is there a way to keep them straight so I don't mix them up?

A: There are a couple of tricks that might help you. First of all, if you can remember that circumference is also found from πD then you can compare that and find which one matches ($2\pi r$). Also, if you can remember that you learned about circumference before area, the 2 comes first on the circumference formula, and comes after in the area formula.

Q: What if I have to find just the area of a slice of pizza, and not the whole circle?

A: Sector area is easy to find—just like you used its angle to find the length of an arc, you can use the sector angle to find sector area. Find the area of the circle then divide it by 360 (to find the area of a 1 degree sector) and multiply by the angle of your sector.

Mario's pizza is here to stay

That was some pretty impressive work—finding the formula for the area of a circle is no small deal, and for Mario it's a huge win in the pizza wars.

From: MegaSlice CEO
To: You
cc: MegaSlice Marketing, MegaSlice Legal

After careful consideration of your "pizza area" calculations, we have withdrawn the TV ad you were referring to and have decided to relocate this store.

Sincerely, MegaSlice CEO

I've expanded into their shop, too! Business couldn't be better. I can't thank you enough—free pizza for life?

As part of a balanced diet of course....

CHAPTER 5

Your Geometry Toolbox

You've got Chapter 5 under your belt and now you've added properties of triangles to your toolbox. For a complete list of tool tips in the book, head over to www.headfirstlabs.com/geometry.

$$\text{Circumference} = \pi D = 2\pi r$$

$$\text{Area} = \pi r^2$$

$$\text{Radius} = \frac{\text{diameter}}{2}$$

$$\text{Sector area} = \pi r^2 \times \frac{\text{Sector angle}}{360}$$

$$\text{Arc length} = 2\pi r \times \frac{\text{Sector angle}}{360}$$

6 quadrilaterals

It's hip to be square

Dang it, I should stick to geometry. I never get my kites mixed up in the quadrilateral family tree.

Maybe three isn't the (only) magic number.

The world isn't just made up of triangles and circles. Wherever you look, you'll see **quadrilaterals**, shapes with four straight sides. Knowing your way 'round the quad family can save you a lot of time and effort. Whether it's **area**, **perimeter**, or **angles** you're after, there are *shortcuts galore* that you can **use to your advantage**. Keep reading, and we'll give you the lowdown.

Edward's Lawn Service needs your help

Edward runs his own lawn mowing and edging service, and over the summer months, demand is high. He's offered you a job to help him cope with all the extra business.

Edward charges clients based on lawn area. That way, clients with a small area of lawn get charged a lower amount than clients with a large lawn.

MMMMMMMMEdward's Lawn Service MMMMMMM

Cost per square meter - $0.10

(Payable weekly. Cost includes one lawn cut per week, with all lawn edges trimmed and neatened.)

I've got more clients than I can handle! I've got a mower you can use, you just need to be able to figure out what to charge people. No problem, right?

Edward →

Your first lawn

The first client Edward hands over to you has a not-so-square
lawn. How much should you charge for mowing it?

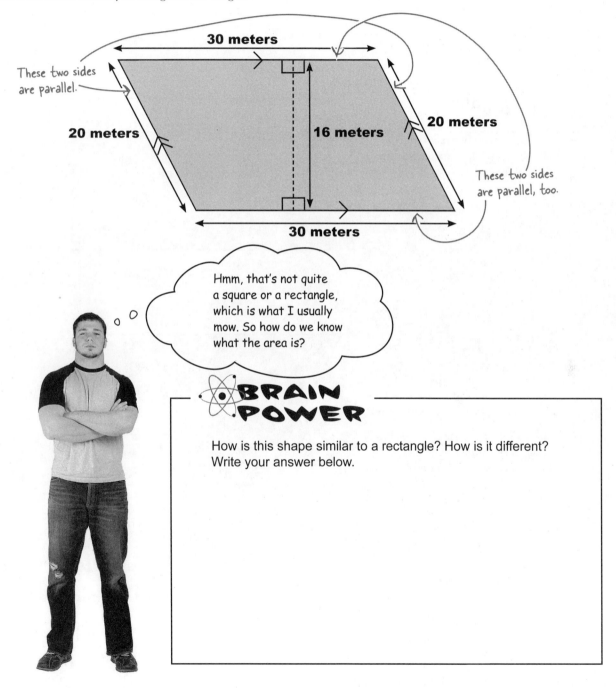

These two sides are parallel.

30 meters

20 meters

16 meters

20 meters

These two sides are parallel, too.

30 meters

Hmm, that's not quite a square or a rectangle, which is what I usually mow. So how do we know what the area is?

BRAIN POWER

How is this shape similar to a rectangle? How is it different?
Write your answer below.

The lawn is a parallelogram

It kinda looks like a rectangle, but squished down so the sides are slanted. A ***parallelogram*** is a four-sided shape whose opposite sides are parallel to each other.

A rectangle:

Rectangles have two pairs of parallel sides. The parallel sides are also congruent.

Each of the four corners forms a right angle.

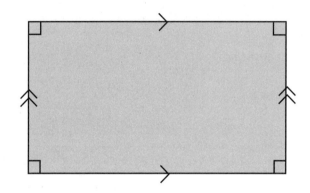

A parallelogram:

Parallelograms have two pairs of parallel sides, too, and like a rectangle, the parallel sides are also congruent.

Unlike a rectangle, the corners of a parallelogram aren't right angles. The most we can say is that opposite corner angles are congruent.

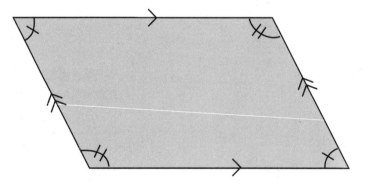

So what should you charge?

We need to calculate the charge for mowing a lawn in the shape of a parallelogram, and to do this, we need to work out the area. You know how to do that for a rectangle, but would that work for a parallelogram, too?

A parallelogram is a four-sided shape whose opposite sides are parallel.

Let's split the parallelogram

Think back to your work in previous chapters. When you needed to find the total area of a shape you were unfamiliar with, you split your shape up into other simpler shapes you were more familiar with.

If you look carefully, you'll see that the parallelogram is actually made up of a central rectangle, with a right angle triangle on either end. This means that to find the area of the parallelogram, we can find the areas of these shapes, and then add them up together.

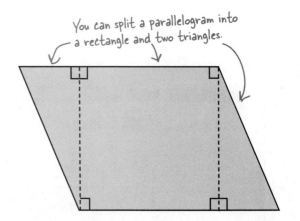

You can split a parallelogram into a rectangle and two triangles.

Sharpen your pencil

What's the total area of the lawn that Edward has asked you to mow? How much should you charge? We've added on some extra measurements to get you started.

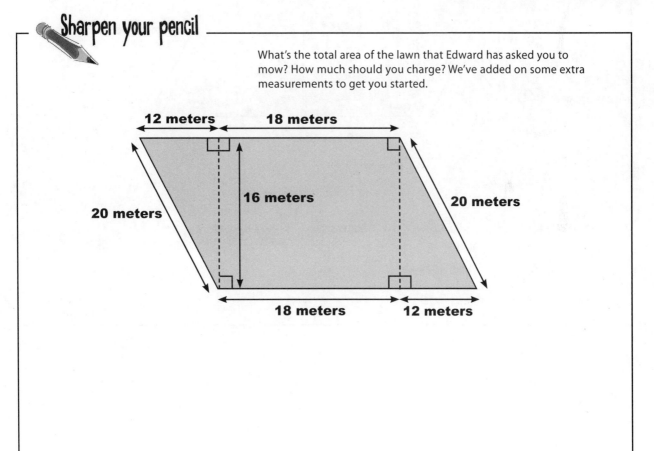

Sharpen your pencil
Solution

What's the total area of the lawn that Edward has asked you to mow? How much should you charge?

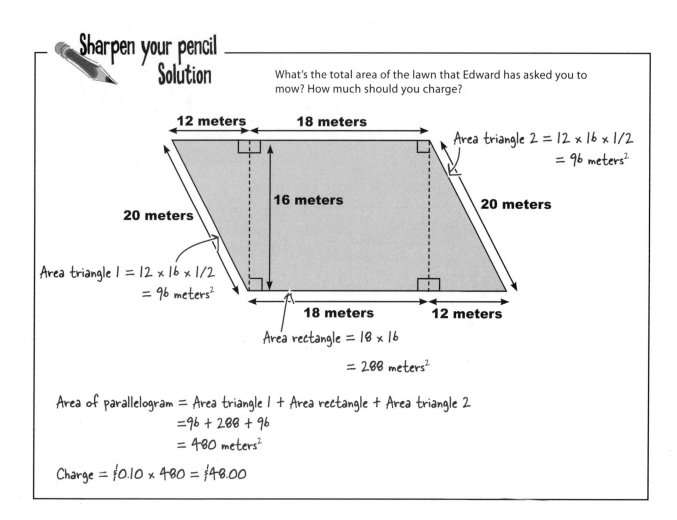

12 meters   18 meters

Area triangle 2 = 12 × 16 × 1/2
= 96 meters²

16 meters

20 meters   20 meters

Area triangle 1 = 12 × 16 × 1/2
= 96 meters²

18 meters   12 meters

Area rectangle = 18 × 16

= 288 meters²

Area of parallelogram = Area triangle 1 + Area rectangle + Area triangle 2
= 96 + 288 + 96
= 480 meters²

Charge = $0.10 × 480 = $48.00

Business is booming!

Before too long, lots of new customers come flocking to the business.

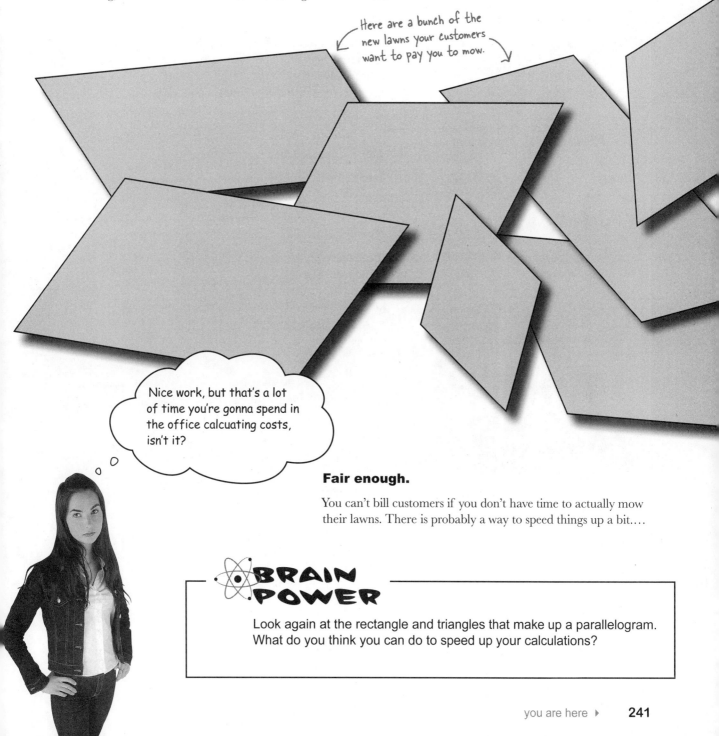

Here are a bunch of the new lawns your customers want to pay you to mow.

Nice work, but that's a lot of time you're gonna spend in the office calcuating costs, isn't it?

Fair enough.

You can't bill customers if you don't have time to actually mow their lawns. There is probably a way to speed things up a bit....

⚛️ BRAIN POWER

Look again at the rectangle and triangles that make up a parallelogram. What do you think you can do to speed up your calculations?

So how can we speed up the area calculation for a parallelogram?

Jim: Well, a parallelogram is basically made up of just three shapes, a rectangle and two triangles.

Frank: Yeah, but those two triangles look the same, right? So maybe they're congruent....

Joe: Nice! So we could calculate one triangle area, multiply by two, and then add on the area of the rectangle.

Jim: That's still a few calculations per lawn, duh!

Frank: True. So I wonder if we could take it any further?

Joe: Maybe we can move these shapes around a bit. Could we turn them into just one shape whose area we know how to calculate?

Frank   Jim   Joe

Geometry Magnets

Let's see if we can figure out a quicker way of calculating the area of a parallelogram. First, arrange the shapes below to form a parallelogram. Then, see if you can rearrange them to form a rectangle. What does this tell you about a more general formula for finding the area of *any* parallelogram?

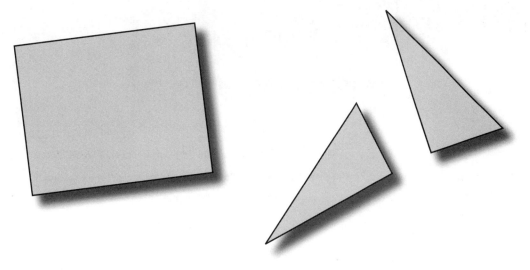

Geometry Magnets Solution

Let's see if we can figure out a quicker way of calculating the area of a parallelogram. First, arrange the shapes below to form a parallelogram. Then, see if you can rearrange them to form a rectangle. What does this tell you about a more general formula for finding the area of *any* parallelogram?

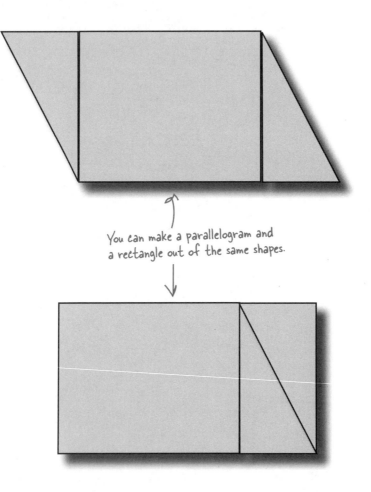

You can make a parallelogram and a rectangle out of the same shapes.

The areas of both the parallelogram and the rectangle must be the same because they're made up of the same shapes. So you can find the area of a parallelogram by finding the area of a rectangle with the same base length and height.

If you don't like what you're given, change it

True, geometry has lots of rules, but moving things around doesn't mean you're breaking them. By moving the shapes you created in the parallelogram around, you were able to form a rectangle. And you know that it has the same height and same base width as the original parallelogram. So now…you know how to find the area of a parallelogram!

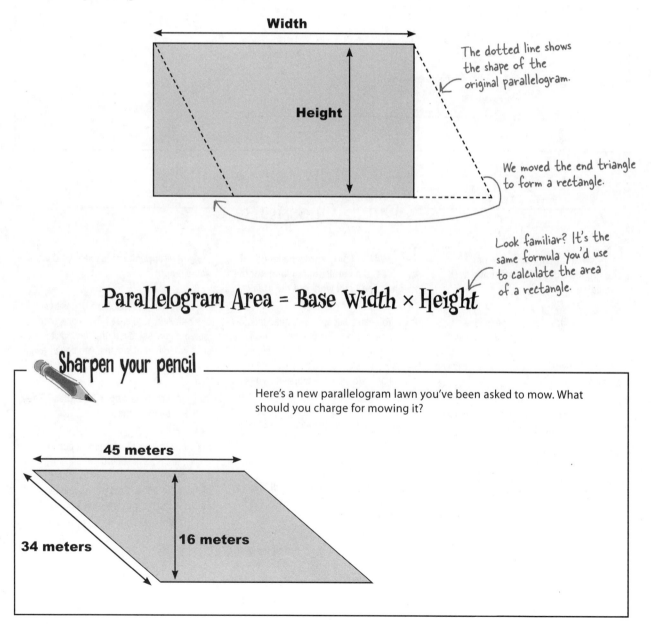

Width

Height

The dotted line shows the shape of the original parallelogram.

We moved the end triangle to form a rectangle.

Parallelogram Area = Base Width × Height

Look familiar? It's the same formula you'd use to calculate the area of a rectangle.

Sharpen your pencil

Here's a new parallelogram lawn you've been asked to mow. What should you charge for mowing it?

45 meters

34 meters

16 meters

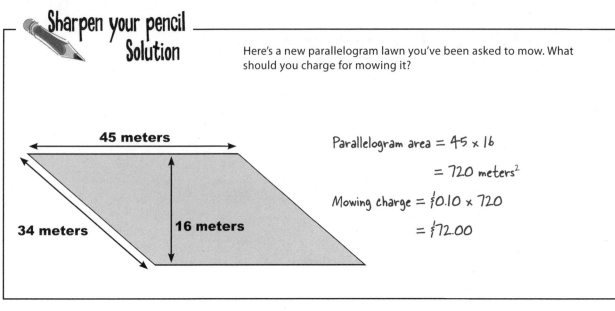

Sharpen your pencil Solution

Here's a new parallelogram lawn you've been asked to mow. What should you charge for mowing it?

45 meters

34 meters

16 meters

Parallelogram area = 45 × 16

= 720 meters²

Mowing charge = $0.10 × 720

= $72.00

---

there are no Dumb Questions

Q: What if I only have one pair of opposite sides? Do I still have a parallelogram?

A: No. In order for a shape to be a parallelogram, both pairs of opposite sides must be parallel. If you only have one pair of opposite parallel sides, your shape isn't a parallelogram.

Q: Can I still use the same formula for finding the area for that shape?

A: This formula is specific to parallelograms. But keep reading, we're going to cover more than just parallelograms and we might have just the formula you need.

Q: What if I only know the length of the sides of a parallelogram and not the height? Can I still work out the area?

A: You can't, because you need to know what the height is. The reason for this is that the degree to which the parallelogram "slants" can make a big difference to the overall area. As an example, if the sides of a parallelogram are tilted so that they're almost horizontal, you'll have a much smaller area than if the sides are almost vertical.

Q: But what if I have extra information such as the internal angles of the parallelogram? Can I work out the area then?

A: Yes, it is possible to find out the area, but it takes a bit more work, not to mention some trigonometry!

Q: Are opposite sides always the same length?

A: Yes they are, because opposite sides of a parallelogram are always congruent (same angle, same length). If opposite sides are different lengths, you don't have a parallelogram.

Q: What about opposite angles? They look the same to me.

A: Yes, for parallelograms opposite angles are always congruent. Similarly, consecutive angles are supplementary. You can use the angle skills you developed earlier in the book to work this out.

But people are upset with Ed's prices...

> You call these charges fair??? I've been timing you. It takes you way less time to mow my lawn than my neighbor's, but we're charged the same amount! If you don't come up with something better, I'll find someone else to do my lawn and tell your other clients that you're up to something....

← Your next customer. Can anyone say too much caffeine?

So what went wrong?

Both customers are charged the same amount, because their lawns must have the same area. But why does one lawn take longer to maintain than the other?

BRAIN POWER

What other factors are involved? Why could it take different lengths of time to maintain the lawns if the lawn areas are the same?

Let's compare the two lawns

They should be the same, but let's see if we can track down what went wrong by comparing the areas of the two lawns. The angry customer has a rectangular lawn, and his neighbor has a lawn in the shape of a parallelogram.

Lawn 1—a rectangle

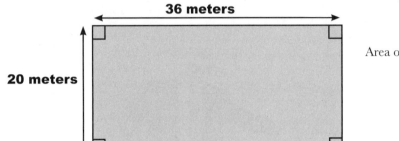

36 meters

20 meters

Area of rectangle = 36 × 20

= 720 meters²

Lawn 2—a parallelogram

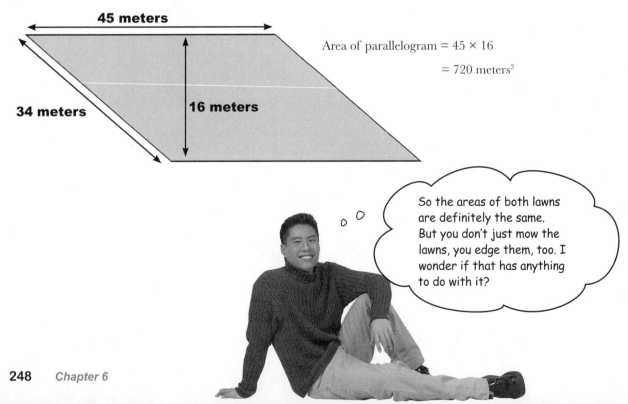

45 meters

34 meters

16 meters

Area of parallelogram = 45 × 16

= 720 meters²

So the areas of both lawns are definitely the same. But you don't just mow the lawns, you edge them, too. I wonder if that has anything to do with it?

The lawns need edging, too

Of course! When you mow lawns for Edward, you have to use an edger to trim where the grass ends, around the outside of each lawn.

The edge of the lawn is the lawn's ***perimeter***. ↖ Remember this from Chapter 4?

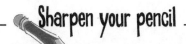

Sharpen your pencil

Calculate the perimeters of the two lawns. How does this account for the difference in time spent on each lawn? What do you think Edward should do?

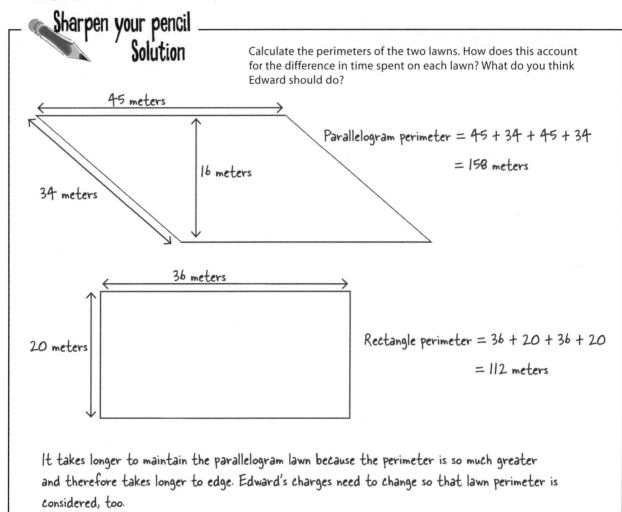

Sharpen your pencil
Solution

Calculate the perimeters of the two lawns. How does this account for the difference in time spent on each lawn? What do you think Edward should do?

45 meters

34 meters

16 meters

Parallelogram perimeter = 45 + 34 + 45 + 34
= 158 meters

36 meters

20 meters

Rectangle perimeter = 36 + 20 + 36 + 20
= 112 meters

It takes longer to maintain the parallelogram lawn because the perimeter is so much greater and therefore takes longer to edge. Edward's charges need to change so that lawn perimeter is considered, too.

Same shape, different perimeters

Your friends ran into a similar problem back in Chapter 4 when they were picking out a venue for the rock festival—the biggest field didn't necessarily have the most perimeter. For our lawn calculations, even if we know the area of a particular lawn, we can't make any assumptions about its perimeter or how long it will take to edge the lawn. And currently, Edward's charges are only based on lawn area.

Same Area ≠ Same Perimeter

there are no
Dumb Questions

Q: So it's not just triangles that have perimeter?

A: No, not at all. The perimeter of a shape is basically the edge around the outside of a shape, whether it's a triangle, rectangle, parallelogram, or some other shape.

Q: Are there any shortcuts we can take in calculating the perimeter of a particular shape?

A: It all depends on the shape.

You calculate the perimeter of a shape by adding the lengths of all its sides together. You can take a few shortcuts if some of these sides are the same length.

As an example, a square has four sides that are all the same length, so the perimeter of a square is just 4 times the length of one side.

Q: What about for a parallelogram?

A: A parallelogram has two pairs of congruent sides, so there are two unique side lengths. You can find the perimeter of a parallelogram by adding these two unique side lengths together, and then multiplying by 2.

Q: What if two shapes have the same area and the same perimeter? Does this mean they're the same shape?

A: Not automatically, no, you need more information. Many different shapes can have the same area and the same perimeter.

Q: More information? Like what?

A: Well, a good starting point is if you know what sort of shapes you're dealing with. This will tell you how many sides your shapes have, and how many of these are congruent.

As an example, if you're told you have two squares and they both have the same area, the squares must be congruent. This is because you calculate the area of a square by calculating (side length)^2. The only way the squares can have the same area is if their sides are the same length.

In general, if you know what type of shape you're dealing with, you can use the properties of the shape to help you deduce further facts.

BULLET POINTS

- A parallelogram is a four-sided shape whose opposite sides are parallel to each other.

- Opposite sides of a parallelogram are congruent.

- Opposite angles of a parallelogram are congruent.

- Consecutive angles of a parallelogram are supplementary.

- To calculate parallelogram area, multiply the width by the height:

- To calculate parallelogram perimeter, add together the length of each side.

Edward changed his rates...

Edward was psyched that you figured out what was going on, and he's changed his pricing to include both the area *and* perimeter of the lawn.

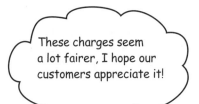

These charges seem a lot fairer, I hope our customers appreciate it!

Edward's Lawn Service

Lawn cutting cost - $0.05 per square meter

Lawn edging cost - $0.10 per meter

(Payable weekly)

...and the customers keep flooding in

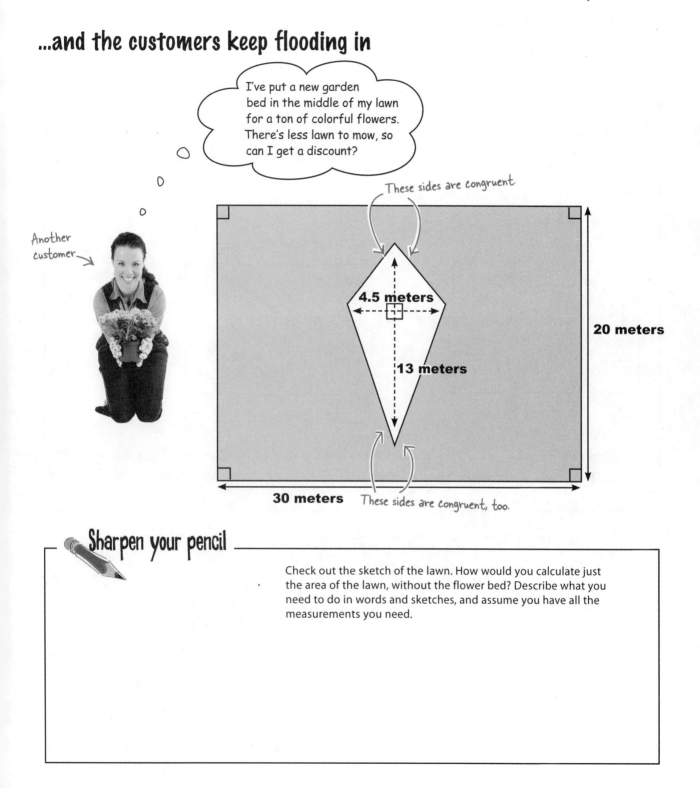

I've put a new garden bed in the middle of my lawn for a ton of colorful flowers. There's less lawn to mow, so can I get a discount?

These sides are congruent.

Another customer

4.5 meters

20 meters

13 meters

30 meters   These sides are congruent, too.

Sharpen your pencil

Check out the sketch of the lawn. How would you calculate just the area of the lawn, without the flower bed? Describe what you need to do in words and sketches, and assume you have all the measurements you need.

Sharpen your pencil
Solution

Check out the sketch of the lawn. How would you calculate just the area of the lawn, without the flower bed? Describe what you need to do in words and sketches, and assume you have all the measurements you need.

Area of lawn = Area of the main lawn rectangle − Area of the flower bed

OR

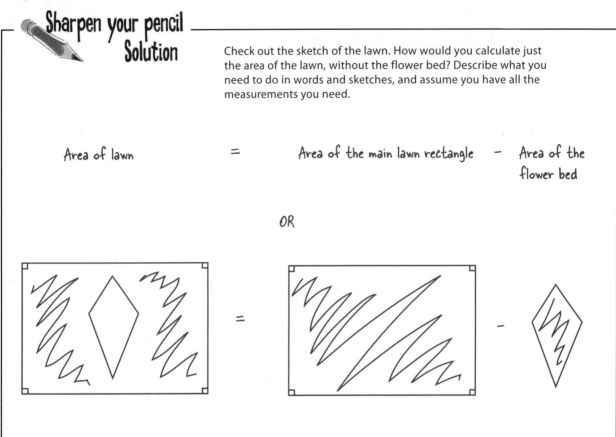

So to find the total area of the lawn we have to mow, we need to find the area of the lawn as though the flower bed wasn't there, and then subtract the area of the flower bed.

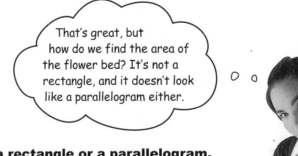

That's great, but how do we find the area of the flower bed? It's not a rectangle, and it doesn't look like a parallelogram either.

True, it isn't a rectangle or a parallelogram.

The opposite sides of the flower bed aren't parallel. Could you use a technique from earlier in this chapter to try to figure out the area of that shape?

Geometry Magnets

Arrange the triangles below to form the same shape as the flower bed. Then see if you can rearrange them to form a rectangle. What's the base of this rectangle? What's the height? What does this tell you about how you might find the area of the flower bed?

Hint: you'll need to flip some of the shapes over to get the rectangle.

Geometry Magnets Solution

Rearrange the triangles below to form the same shape as the flower bed. Then see if you can rearrange them to form a rectangle. What's the base of this rectangle? What's the height? What does this tell you about how you might find the area of the flower bed?

Hint: you'll need to flip some of the shapes over to get the rectangle.

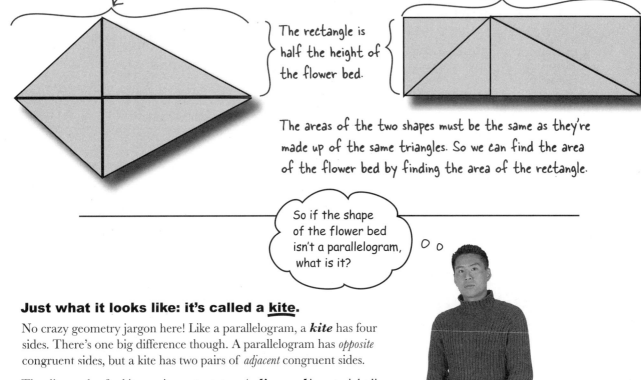

The rectangle and flower bed have the same width.

The rectangle is half the height of the flower bed.

The areas of the two shapes must be the same as they're made up of the same triangles. So we can find the area of the flower bed by finding the area of the rectangle.

So if the shape of the flower bed isn't a parallelogram, what is it?

Just what it looks like: it's called a <u>kite</u>.

No crazy geometry jargon here! Like a parallelogram, a ***kite*** has four sides. There's one big difference though. A parallelogram has *opposite* congruent sides, but a kite has two pairs of *adjacent* congruent sides.

The diagonals of a kite are important, too. A ***diagonal*** is a straight line that connects one corner to the corner opposite. For a kite, these diagonals are always perpendicular to each other.

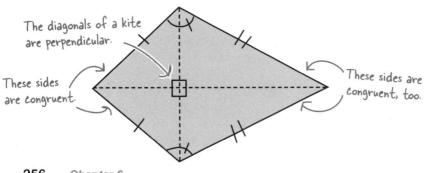

The diagonals of a kite are perpendicular.

These sides are congruent.

These sides are congruent, too.

Use diagonals to find the area of the kite

Just like the parallelogram, there's a shortcut we can take to find the area of a kite. All we need to do is multiply the lengths of the two diagonals together, and divide by 2:

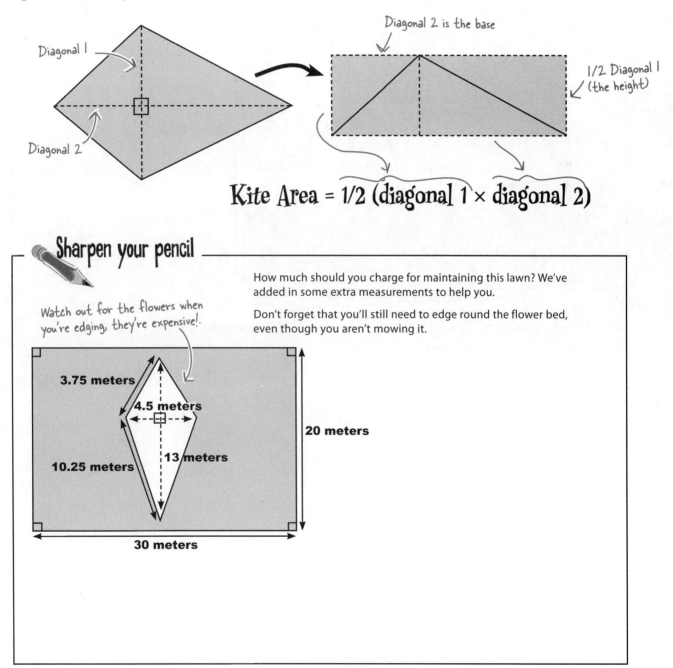

Diagonal 1

Diagonal 2

Diagonal 2 is the base

1/2 Diagonal 1 (the height)

Kite Area = 1/2 (diagonal 1 × diagonal 2)

Sharpen your pencil

How much should you charge for maintaining this lawn? We've added in some extra measurements to help you.

Don't forget that you'll still need to edge round the flower bed, even though you aren't mowing it.

Watch out for the flowers when you're edging, they're expensive!

3.75 meters

4.5 meters

10.25 meters

13 meters

20 meters

30 meters

Sharpen your pencil
Solution

How much should you charge for maintaining the lawn? We've added in some extra measurements to help you.

Don't forget that you'll still need to edge round the flower bed, even though you aren't mowing it.

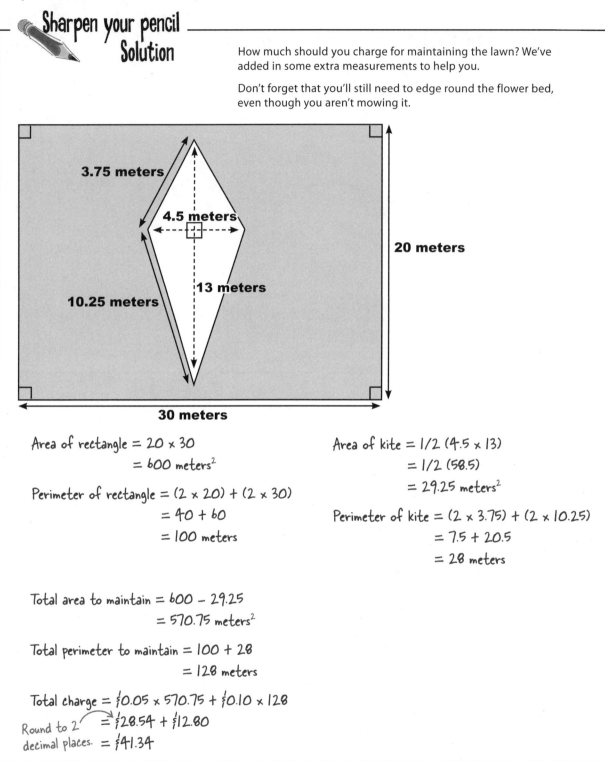

Area of rectangle = 20 x 30
= 600 meters²

Perimeter of rectangle = (2 x 20) + (2 x 30)
= 40 + 60
= 100 meters

Area of kite = 1/2 (4.5 x 13)
= 1/2 (58.5)
= 29.25 meters²

Perimeter of kite = (2 x 3.75) + (2 x 10.25)
= 7.5 + 20.5
= 28 meters

Total area to maintain = 600 − 29.25
= 570.75 meters²

Total perimeter to maintain = 100 + 28
= 128 meters

Total charge = $0.05 x 570.75 + $0.10 x 128
Round to 2   = $28.54 + $12.80
decimal places. = $41.34

Diagonals

The diagonals of a kite are perpendicular, but that's not all there is to say about them.

For starters, one diagonal bisects the other, meaning it chops the other diagonal in half. It also bisects the pair of opposite angles, and if you look at the remaining pair of angles, they're congruent, too. So there's a lot you can know about a kite without having to do any calculations!

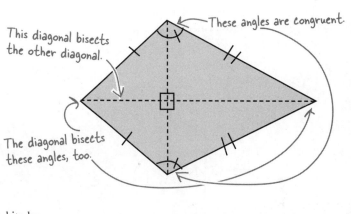

This diagonal bisects the other diagonal.

These angles are congruent.

The diagonal bisects these angles, too.

Area and perimeter

As you discovered earlier, you find the area of a kite by multiplying together the lengths of the two diagonals and dividing by two. To find the perimeter, remember that there are two pairs of congruent sides so you only have to add the two different sides together and multiply by two.

there are no Dumb Questions

Q: So the diagonals of a kite are perpendicular. What about the diagonals of a parallelogram, are they perpendicular, too?

A: In general, parallelograms don't have perpendicular diagonals.

The diagonals of a parallelogram are still important though. If a shape is a parallelogram, then its diagonals bisect each other. Try adding diagonals to the parallelograms earlier in the chapter and you'll see what we mean.

Q: Could I have calculated the area of the kite by splitting it into simpler shapes like before?

A: You could, but it would have taken you much longer to calculate. All you really need to do is multiply the two diagonals together and divide the result by 2.

Q: The kites we've looked at in this chapter look symmetrical. Is that a coincidence?

A: No, not at all. Every kite is symmetrical along one diagonal.

Q: Can a shape be both a parallelogram and a kite?

A: Yes it can. A shape is a parallelogram and a kite if it fits the description of both. In other words, it must have two pairs of separate adjacent congruent sides, and also the opposing sides must be parallel. This means that all four sides must be congruent.

An example of a shape that is both a parallelogram and a kite is a square. All four sides are congruent, and opposite sides are parallel. We'll get to that in a little bit....

Landowners, unite

Just when all the lawns seemed under control, Edward ran into a little snag. He turned up at a customer's house to find that he has been buying up adjacent land. The neat rectangular lawn that Edward had been mowing for months has been transformed into…something else.

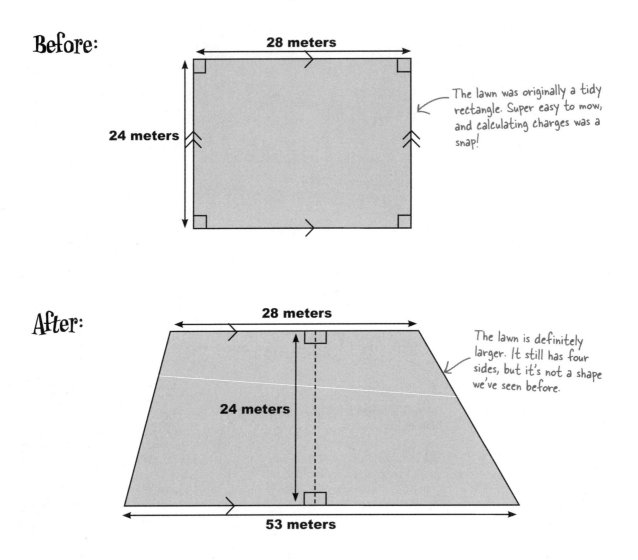

Before:

28 meters

24 meters

The lawn was originally a tidy rectangle. Super easy to mow, and calculating charges was a snap!

After:

28 meters

24 meters

53 meters

The lawn is definitely larger. It still has four sides, but it's not a shape we've seen before.

It's going to take longer to mow and edge this new lawn, and Edward needs a hand working out the charges. He took some rough measurements and gave them to you over the phone, but he didn't get all the sides. So what do we do with a lawn like this?

Sharpen your pencil

Check out the shape of the new lawn. How do you think we could go about finding the area of a shape like this?

> You've **GOT** to be kidding! This is way too easy, we've done all this before. All we need to do is split the shape into a rectangle and two triangles, and find the total area. Duh!

Really?

We can't do exactly the same thing we did before because we don't have all the measurements we'd need. Let's take a closer look.

There are some familiar things about this shape

It's a four-sided shape, but it only has one pair of parallel sides. These parallel sides are called **bases**. We call this shape a **trapezoid.**

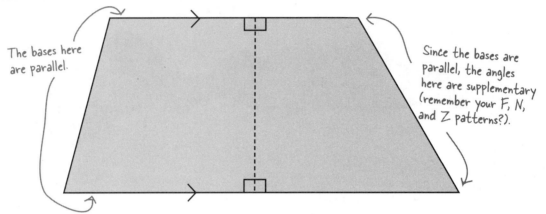

The bases here are parallel.

Since the bases are parallel, the angles here are supplementary (remember your F, N, and Z patterns?).

But what's the area?

In theory, we could split the trapezoid into a rectangle and two right triangles, and then add together the area of each shape. The trouble is, we don't have enough information to do this. We know that they're both right triangles, but we don't know what the other two angles are, and we only know the length of one of the sides. So what can we do instead?

A trapezoid has exactly one pair of opposite parallel sides.

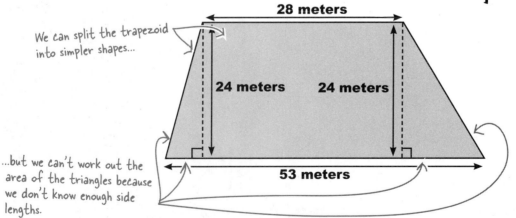

We can split the trapezoid into simpler shapes...

28 meters

24 meters   24 meters

53 meters

...but we can't work out the area of the triangles because we don't know enough side lengths.

Sharpen your pencil

Instead of splitting the shape up, let's add to it and see if that helps us figure out how to find the area of a trapezoid. Here are two congruent trapezoids. Draw them together so that they form a single parallelogram.

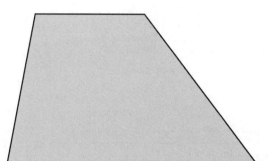 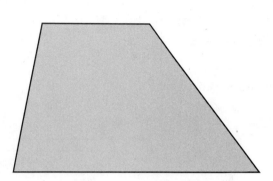

What's the width of the parallelogram?

What's the height of the parallelogram?

What ideas does this give you about how to calculate the area of a trapezoid?

two trapezoids make a parallelogram

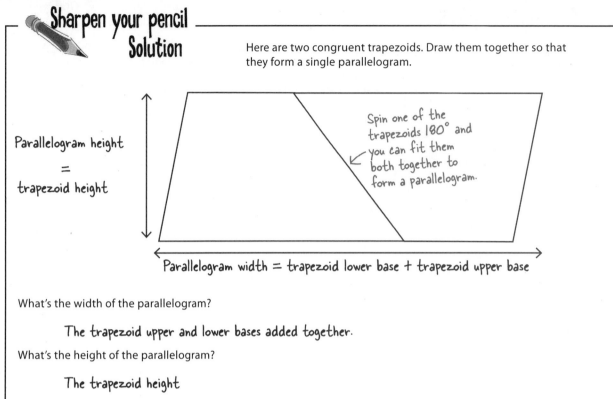

Here are two congruent trapezoids. Draw them together so that they form a single parallelogram.

Parallelogram height
=
trapezoid height

Spin one of the trapezoids 180° and you can fit them both together to form a parallelogram.

Parallelogram width = trapezoid lower base + trapezoid upper base

What's the width of the parallelogram?

The trapezoid upper and lower bases added together.

What's the height of the parallelogram?

The trapezoid height

What ideas do you have about how to calculate the area of a trapezoid?

The parallelogram is made up of two trapezoids. This means that the area of each trapezoid must be half that of the parallelogram, and we know that the area of a parallelogram is equal to height × width.

Calculate trapezoid area using base length and height

So what we've discovered is that, like with the other shapes you've encountered so far, there's a shortcut we can take if we want to find the area of a trapezoid. All we need to do is add together the upper and lower base lengths, multiply by the height, and then divide by two.

$$\text{Trapezoid Area} = \frac{\text{height (base 1 + base 2)}}{2}$$

Exercise

What's the charge for maintaining the trapezoid lawn? Thankfully, Ed went back and collected some extra measurements to help you out with the perimeter.

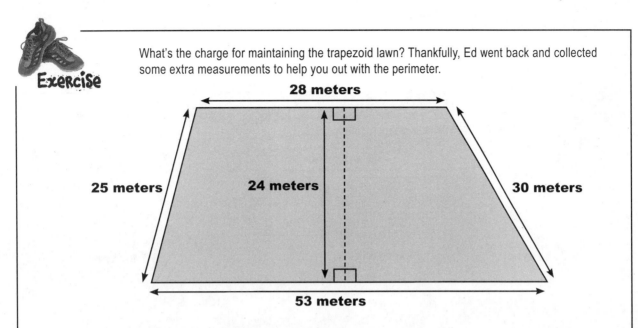

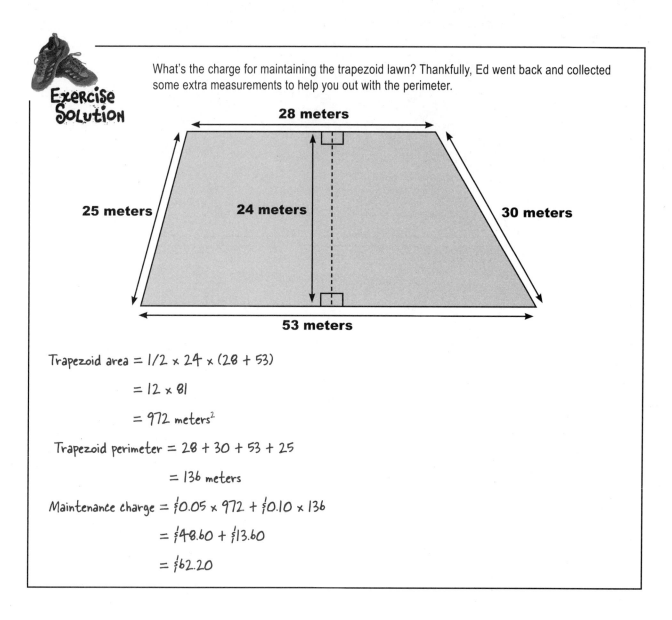

What's the charge for maintaining the trapezoid lawn? Thankfully, Ed went back and collected some extra measurements to help you out with the perimeter.

Trapezoid area = 1/2 × 24 × (28 + 53)

        = 12 × 81

        = 972 meters²

Trapezoid perimeter = 28 + 30 + 53 + 25

        = 136 meters

Maintenance charge = $0.05 × 972 + $0.10 × 136

        = $48.60 + $13.60

        = $62.20

The Trapezoid Exposed

**This week's interview:
What's So Special About Trapezoids?**

Head First: Hey, Trapezoid, nice to have you on the show tonight.

Trapezoid: It's a real pleasure. I don't seem to get out as much as some of the other shapes.

Head First: What do you mean?

Trapezoid: Well, everyone knows about Square and Rectangle, although they're kind of boring if you ask me. Kite gets good press because it looks like… well…a kite. But me? Everyone forgets about me.

Head First: I notice that you didn't mention Parallelogram. Don't you guys have something in common?

Trapezoid: Some would say we're family, and I guess you could say we're kinda like cousins. I have only one pair of parallel sides, and Parallelogram has two. But don't you think that's just being greedy? Parallelogram's just a lopsided, funny-looking Rectangle if you ask me, and I don't limit myself like that.

Head First: But aren't you just like Triangle but with one of the points chopped off?

Trapezoid: That's one way you could talk about me, yes, but the key thing is that I have one pair of parallel sides, and Triangle doesn't. Um, I wouldn't talk to Triangle about that, he's still a bit sensitive.

Head First: But don't you have one nice thing in common with Triangle?

Trapezoid: You did your research, didn't you? We've all heard of an Isosceles Triangle, right? Well, Trapezoids can be Isosceles, too.

Head First: What's special about that?

Trapezoid: It's still a Trapezoid, but with congruent legs. And by legs I mean the sides that aren't bases. Isosceles Trapezoid is a bit more regular than me, there's more symmetry about him. He's thinking about getting into modeling, from what I hear. At the very least, it must make shaving much easier.

Head First: What about diagonals?

Trapezoid: What about them? Sure, I have them, but I can't say much about them. Now Isosceles Trapezoid, he has congruent diagonals, and he has some congruent angles, too. Some guys have all the luck.

Head First: Well, that's all we have time for, Trapezoid, thanks for stopping by.

BULLET POINTS

- A trapezoid is a four-sided shape with one pair of parallel sides called bases.

- Since the bases are parallel, this means that you have two pairs of supplementary angles.

- An isosceles trapezoid is a trapezoid with congruent "legs." The lower base angles are congruent, the upper base angles are congruent, and the diagonals are congruent, too.

The quadrilateral family tree

There's one key thing that the shapes in this chapter all have in common—they all have four sides.

Any shape that has four straight sides is a quadrilateral. Parallelograms, kites, and trapezoids are all part of the quadrilateral family, along with shapes such as squares and rectangles. Here's the quadrilateral family tree so you can see how they're all related.

The arrows show you the hierarchy

The thick arrows on the family tree show you the relationships between the shapes. Any shape at the end of an arrow is a more specialized form of the shape that the arrow is coming from. So a kite is a type of quadrilateral since it shares the same properties as a quadrilateral—it has four sides, with two pairs of separate adjacent sides. Similarly, a square is type of rhombus, which means it is also a type of kite.

The family tree also shows you which shapes are *not* directly related. That means a kite is not a type of parallelogram and a rectangle is not a type of trapezoid.

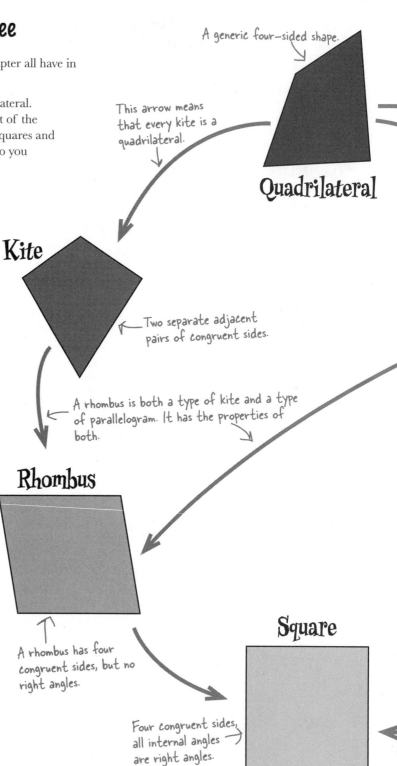

A generic four-sided shape.

Quadrilateral

This arrow means that every kite is a quadrilateral.

Kite

Two separate adjacent pairs of congruent sides.

A rhombus is both a type of kite and a type of parallelogram. It has the properties of both.

Rhombus

A rhombus has four congruent sides, but no right angles.

Square

Four congruent sides, all internal angles are right angles.

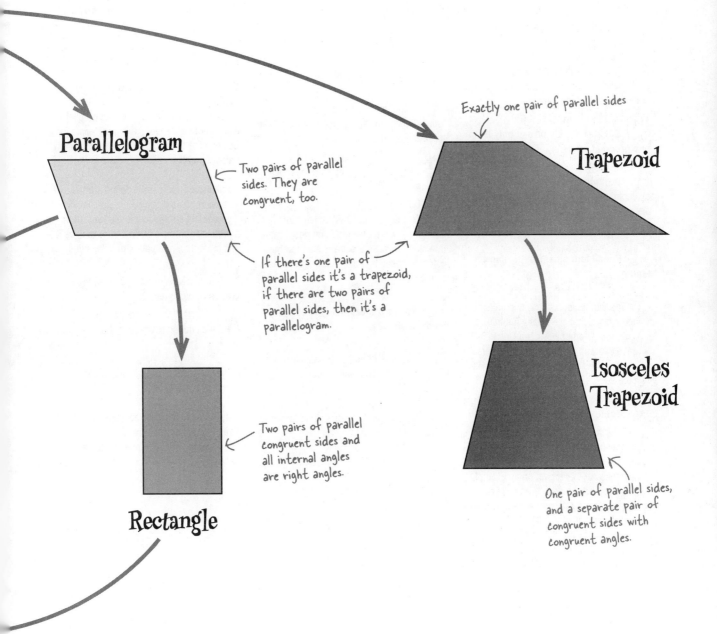

Parallelogram

Two pairs of parallel sides. They are congruent, too.

Exactly one pair of parallel sides

Trapezoid

If there's one pair of parallel sides it's a trapezoid, if there are two pairs of parallel sides, then it's a parallelogram.

Two pairs of parallel congruent sides and all internal angles are right angles.

Rectangle

Isosceles Trapezoid

One pair of parallel sides, and a separate pair of congruent sides with congruent angles.

there are no
Dumb Questions

Q: So how does knowing which different quadrilaterals relate to each other help me?

A: Some shapes can be classified as more than one type of quadrilateral, and if you know which groups each shape belongs to, you can use the properties of each group to help you calculate things like area and perimeter.

Q: But surely a shape can only be in one group at a time?

A: No, it's really a hierarchy. As an example, a square isn't just a square, it's also a kite as it has two pairs of adjacent sides.

Q: But all four sides are the same length.

A: Yes, but because it has two pairs of adjacent sides that are equal, that makes it a kite, too. Another way of looking at it is that a kite with four right angle corners where the sides are all the same length is called a square.

Q: But how does this really help me?

A: If a shape belongs to a group then you can find the area and perimeter using the formulas for that group. As an example, suppose you have a square and all you know about it is the length of the diagonals. How would you go about finding the area? One approach would be to calculate the sides of the square using the length of the diagonal. The area would then be the square of the side length. A simpler approach, however, would be to remember that a square is also a kite, and use the kite formula to find the area. In other words, multiply the diagonals together and divide by 2.

Q: Oh, I get it. If I can remember how all the quadrilaterals relate to one another, I can save time calculating things like area.

A: Absolutely!

Q: So what's a rhombus?

A: A rhombus is quadrilateral where all four sides are the same length. It's a bit like a skewed square.

Q: Oh, I see. So does that mean that a square is a rhombus, too?

A: Well done! Yes, that's totally right. But remember, the reverse isnt necessarily true. Just because a square is a rhombus, it doesn't mean that every rhombus is also a square.

Q: So is a trapezoid a type of parallelogram?

A: A parallelogram has to have two pairs of parallel sides, and a trapezoid has to have exactly one pair of parallel sides. This means that a trapezoid is not a parallelogram, as trapezoids don't have two pairs of parallel sides.

Q: The family tree shows that a parallelogram is not a type of trapezoid. In another book, I saw a diagram where the two are shown as related. Why's that?

A: That's an interesting question.

We've defined a trapezoid as having *exactly* one pair of parallel sides. Since a parallelogram has *two* pairs of parallel sides, it doesn't meet the requirements for being in the trapezoid club. It has too many pairs of parallel sides.

Sometimes other people define a trapezoid as having *at least* one pair of parallel sides. If you define a trapezoid in this way, then a parallelogram is a type of trapezoid, as it has more than one pair of parallel sides.

Q: I've heard of a trapezium, too. What's one of those?

A: It all depends where in the world you are.

In the U.S., a trapezium is a quadrilateral with no parallel sides.

Outside the U.S., a trapezium is the name given to a shape with one pair of parallel sides—what people in the U.S. call a trapezoid.

You've entered the big league

Thanks to your skill with quadrilaterals, Ed's lawn mowing business has really blown up.

> We're raking the money in now, and it's all thanks to you. We've even been approached by the local fancy golf course for a multi-year contract. Want to be my business partner?

That means a much bigger cut of the profits for you!

Quadville
Country Club

CHAPTER 6

Your Geometry Toolbox

You've got Chapter 6 under your belt and now you've added quadrilaterals to your toolbox. For a complete list of tool tips in the book, head over to www.headfirstlabs.com/geometry.

Parallelogram

Has two pairs of parallel sides that are also congruent.

Opposite corner angles are congruent.

Area = height × base width.

Quadrilateral

A flat shape with four straight sides.

Rectangle

Has two pairs of parallel congruent sides.

Every corner is a right angle.

Area = width × height.

Trapezoid

Has one pair of parallel sides called bases.

Area = 1/2 × height × (base 1 + base 2)

Rhombus

Has four congruent sides.

Is both a parallelogram and a kite—so you can use the area calculation of either shape.

Kite

Has two pairs of separate adjacent congruent sides.

Diagonals are perpendicular.

Area = 1/2 × diagonal 1 × diagonal 2

7 regular polygons

It's all shaping up

Would you like to try some? It's made with my secret ingredient: triangles. They're just so versatile—I put them in everything I bake.

Want to have it your way?

Life's full of compromises, but you don't have to be restricted to triangles, squares, and circles. **Regular polygons** give you the **flexibility** to demand exactly the shape you need. But don't think that means learning a list of new formulas: *you can treat 6-, 16-, and 60-sided shapes the same*. So, whether it's for your own creative project, some required homework that's due tomorrow, or the demands of an important client, you'll have the tools to **deliver exactly what you want**.

We need to choose a hot tub

Man, the music festival was such a hit we're doing it over again! Everything's set except one thing: the band wants a hot tub backstage!

Choosing a hot tub? Now, that sounds like a job that should be super easy. But of course there's a catch: everybody's got an opinion on what this hot tub needs to be like…

…and they need it all sorted out by tomorrow.

A bunch of different people each have requirements about this hot tub.

BAND:

Want hot tub which gives the most butt-space.

ENVIRONMENTAL ENGINEER:

Says max 3 cubic meters water!

CARPENTERS:

Need to know what dimensions and angles to cut—by tomorrow at the latest, please.

All the hot tubs are regular polygons

The local hot tub suppliers have said they'll rush through an order for any hot tub in their range—all of which they describe as **regular polygons.** The word **regular** is another way of saying equilateral— all the sides are the same length—but it means the polygons have some other useful properties as well.

A polygon is simply any flat shape with three or more sides. You've already worked with a bunch of them, like triangles and quads.

Sharpen your pencil

What properties of the hot tubs do you need to work out to satisfy:

1. The bands? ....

2. The environment engineer? ....

3. The carpenters? ....

The carpenters need to make an appropriately sized hole in the floor for the tub to go in.

How does it help you that the tubs are all **regular** polygons?

....

....

Sharpen your pencil
Solution

What properties of the hot tubs do you need to work out to satisfy:

1. The bands? .... **Perimeter** .... ⭠ People sit around the edges.

2. The environment engineer? **Volume** ⭠ If you put area here that's cool, too! You'll see why in a bit.

3. The carpenters? .... **Internal angles** ....
.... **Side lengths** .... ⭠ Carpenters want these angles.

How does it help you that the tubs are all *regular* polygons?

Since all the sides on a regular polygon are the same length, then all the angles are equal too, so we've only got to find one side and one angle—all the others will match.

Regular polygons have equal sides and angles

Any six-sided polygon could have six different side lengths and six different angles. But fortunately for us, all the hot tubs are *regular* polygons. A six-sided regular polygon has six sides of equal length and six equal angles.

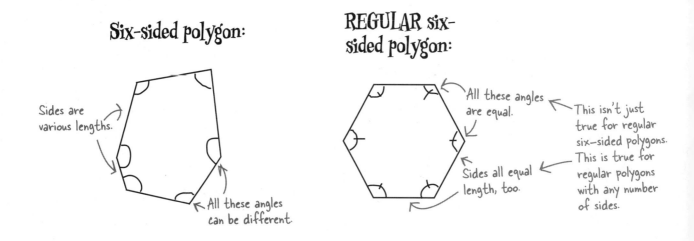

Six-sided polygon:

Sides are various lengths.

All these angles can be different.

REGULAR six-sided polygon:

All these angles are equal.

Sides all equal length, too.

This isn't just true for regular six-sided polygons. This is true for regular polygons with any number of sides.

Butt-space is all about perimeter

The number of people who can sit in the tub at
the same time depends on the perimeter of the tub.
More perimeter = more butt-space = more people.

People sit around
the edges.

More edge means
more space for
people to sit.

So we need the hot tub with the
biggest perimeter? But we've also got
to limit the water...so it's not as simple
as just "bigger is better," is it?

That's true—in fact this hot tub problem is a lot deeper than it first appears.

First, you've got to compare perimeters of polygons which are
all different shapes, and second you've got to work to a limited
amount of water...3 cubic meters.

BRAIN POWER

Is there a pattern you can use to work out
the perimeters of your regular polygon
shaped hot tubs? What would you need
to know first?

Is 3 cubic meters of water a lot or a little?

Typical engineer. Instead of giving you the amount of water in terms of a measure that might actually make some sense, like cups or spoonfuls or gallons, he's talking about ***cubic meters***. So, how much is 3 cubic meters of water? A bathful? A pondful? A swimming poolful?

Let's start by think about what just 1 cubic meter actually is:

Volume of a cube = length x width x depth

Cubic meter = 1m x 1m x 1m

= 1m³

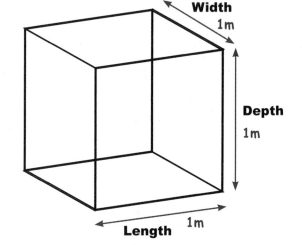

A cube-shaped container with length, width, and depth all of one meter holds exactly one cubic meter of water.

So 3 cubic meters of water is the volume of water you could store in three of those one meter cubes:

 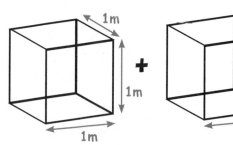 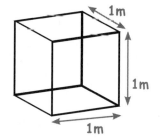

3 cubic meters =

OK—stop right there! I don't know how closely **you've** looked but I don't see any hot tubs that look like they're made out of **CUBES**.

That's true.

In fact, thinking about cube shapes can only take us so far, because, aside from the square-shaped one, the hot tub range isn't made out of cubes.

How do the hot tubs compare to a cube?

Like a cube, all the hot tubs have a uniform depth. They also have straight sides at a right angle to the base. But for most of the shapes of hot tub, calculating the volume using **length × width × depth** would give you the volume of a box that you could fit the hot tub inside, not the water that the hot tub would contain. It might be handy, but it's not what we're looking for!

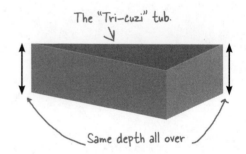

The "Tri-cuzi" tub.

Same depth all over

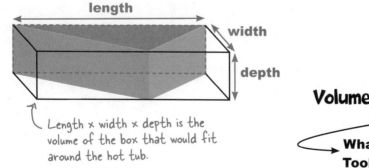

length

width

depth

Length × width × depth is the volume of the box that would fit around the hot tub.

Volume of a hot tub = ? × depth

What tool from your Geometry Toolbox could we use to calculate the volume, instead of "length × width"?

Hot tub volume is <u>area</u> x depth

Calculating the volume of some 3D shapes can get pretty gnarly, but for the hot tub range it's simple. The tubs have straight sides, so the volume can be found from area × depth.

These "straight-sided" shapes are known as prisms—more about those in Head First 3D Geometry.

[Thanks! Marketing xx]

AREA = length x width

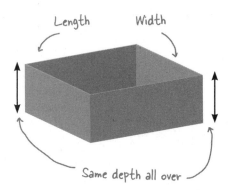

Length   Width

Same depth all over

---

there are no Dumb Questions

Q: This book is called *Head First 2D Geometry*, right? So how come we're talking about 3D in this chapter?

A: Volume is definitely a 3D topic, and we cover it in much more detail in *Head First 3D Geometry*, but it's not too bad to dip your toes in the water is it? Also, we're about to turn this problem back into a 2D one on the next page.

Q: You're going to turn a 3D problem into a 2D problem? How does that work?

A: The third dimension in the hot tubs problem is depth. Once we don't have to work with the depth anymore it's just a 2D problem we're left with. Hold that thought to the bottom of the next page.

Q: What if the tub was deeper at one end than the other? Or had curved sides?

A: The *area × depth* formula only applies to 3D shapes with straight sides, all of the same depth, which are perpendicular to the base. If the hot tub was deeper at one end we'd need a different way of working out its volume.

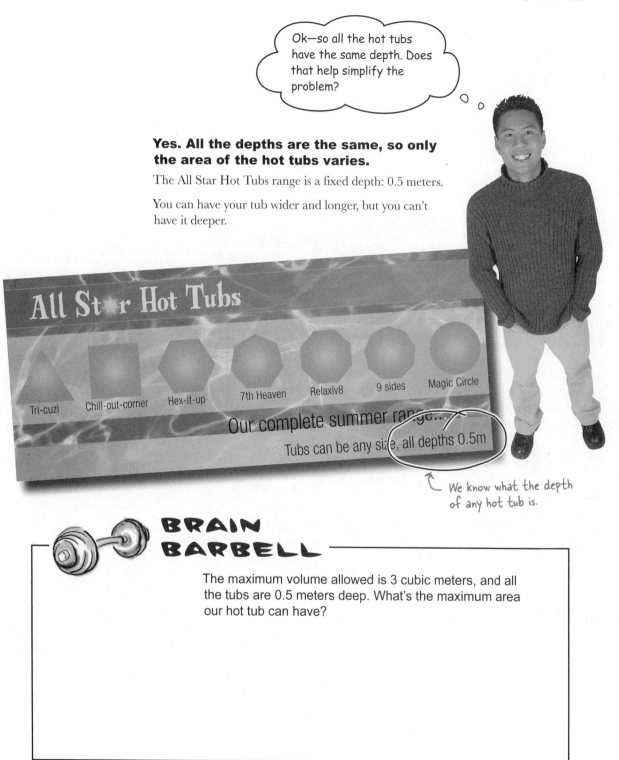

> Ok—so all the hot tubs have the same depth. Does that help simplify the problem?

Yes. All the depths are the same, so only the area of the hot tubs varies.

The All Star Hot Tubs range is a fixed depth: 0.5 meters.

You can have your tub wider and longer, but you can't have it deeper.

All St★r Hot Tubs

Tri-cuzi   Chill-out-corner   Hex-it-up   7th Heaven   Relaxiv8   9 sides   Magic Circle

Our complete summer range...

❋ Tubs can be any size, all depths 0.5m

We know what the depth of any hot tub is.

BRAIN BARBELL

The maximum volume allowed is 3 cubic meters, and all the tubs are 0.5 meters deep. What's the maximum area our hot tub can have?

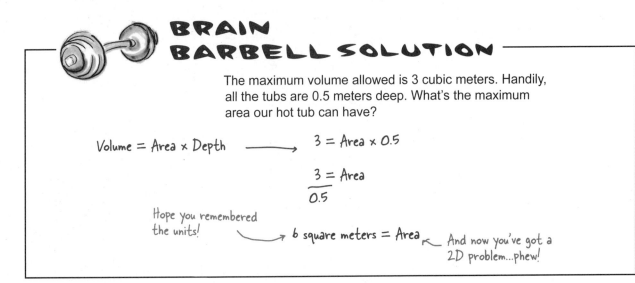

BRAIN BARBELL SOLUTION

The maximum volume allowed is 3 cubic meters. Handily, all the tubs are 0.5 meters deep. What's the maximum area our hot tub can have?

Volume = Area × Depth ⟶ 3 = Area × 0.5

$$3 = \frac{Area}{0.5}$$

Hope you remembered the units!

⟶ 6 square meters = Area ⟵ And now you've got a 2D problem...phew!

The hot tub's area must be 6m²

Technically, anything less than or equal to six would be OK with the engineers, but we want max butt-space, so we want to push it to the limit.

Whichever shape of tub we choose, the area of the hot tub must be no more than 6 square meters. That way we know for sure that it'll only need 3 cubic meters of water to fill it, and the environmental engineers will be happy.

As will we! Saving water is kind of important.

The shape and size of the tub is completely flexible—it's up to us to work out what size and shape of tub will give us 6 square meters of area and also meet the band's requirement: maximum butt-space!

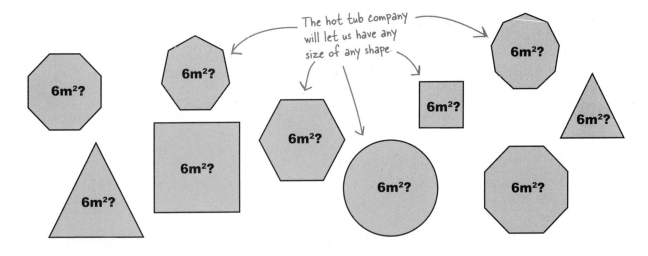

The hot tub company will let us have any size of any shape.

Which hot tub shape gives the most butt-space?

So, will they all have the same perimeter since they've all got the same area?

Jill: No—even for triangles that didn't work—remember?

Joe: Oh, yeah. Well, I guess we just have to divide up the work then.

Frank: What work?

Joe: Well, we'll have to do a bunch of calculations for each shape—like try with the sides at 1 meter, 2 meters, 3 meters, and so on. Working out the area and perimeter for each. Find the ones closest to 6 square meters of area and then compare them and maybe refine the calculations....

Jill: Whoa. No way. That's a ton of work! We'll never get it done for tomorrow.

Frank: I don't think we need to do it that way anyway. Do you think we could use the formulas we know for area, and sort of rewind them to find the side lengths?

Joe: Rewind them? Rewind a formula? What?

Frank

Jill

Joe

What do YOU think Frank means? Would it work?

Work backward from area to find butt-space

Normally, you're given the lengths of the sides of a shape and you push them through a couple of formulas to get area and perimeter:

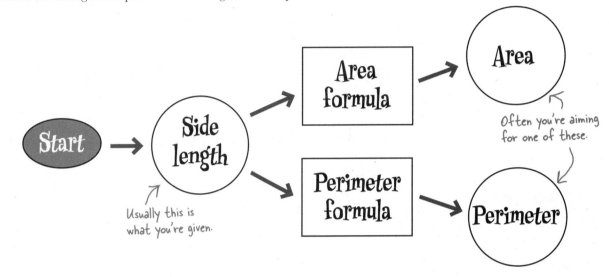

But in this case, you know your hot tub *area*, and you need to find the length of the sides that would give that area, so you can use them to find the perimeter (butt-space). To do this you need to work backward.

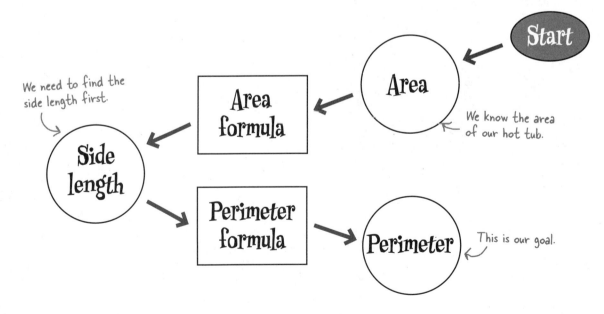

Exercise

Use any formula you know for area of a square to find the side lengths of the square hot tub with 6 square meters area.

Then use the side length to find the perimeter.

If each butt requires 0.5 meters of perimeter, how many butts fit in this square hot tub?

6m²

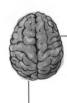

Reviewing isn't cheating! Quads area formulas: page 272.

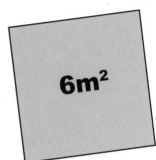

Exercise Solution

Use any formula you know for area of a square to find the side lengths of the square hot tub with 6 square meters area.

Then use the side length to find the perimeter.

If each butt requires 0.5 meters of perimeter, how many butts fit in this square hot tub?

6m²

Area of a square = side length × side length

$b = s^2$

If you had width × height that works too—but in a square the width and height are the same so it's even easier!

$\sqrt{b} = s$

2.45 meters = s ← Length of each side.

Perimeter of a square = 4s

Perimeter = 4 × 2.45 = 9.8 meters

Units help you keep track of what each number means.

Each butt takes 0.5 meters

total butts = $\frac{9.8}{0.5}$ = 19.6 butts

Q: Half a meter per butt? Really?

A: OK, so we all know that butt-size is one of life's interesting variables. Obviously not all butts are the same size, but in this problem that's not too important. Even if you had a bunch of people with more generously proportioned butts, you'd still fit more into the biggest perimeter, so it's still a useful comparison.

Q: 19 butts or 19.6 butts? I've never met anyone with 0.6 of a butt.

A: That's true, you'd really round down to 19. Only whole butts can get in the tub (unless you count a really skinny one as a 0.6). That 0.6 gives a little extra room to the 19 folk in there though, so it's worth keeping those decimals in case we end up choosing between two tubs which both have butt counts between 19 and 20.

Is 19.6 butts a lot or a little?

One number on its own isn't enough to choose a hot tub—you want the MOST possible butt-space for your 6 square meters of area. You need to tackle the other hot tubs shapes to find out if they're better or worse than the square tub.

Sharpen your pencil

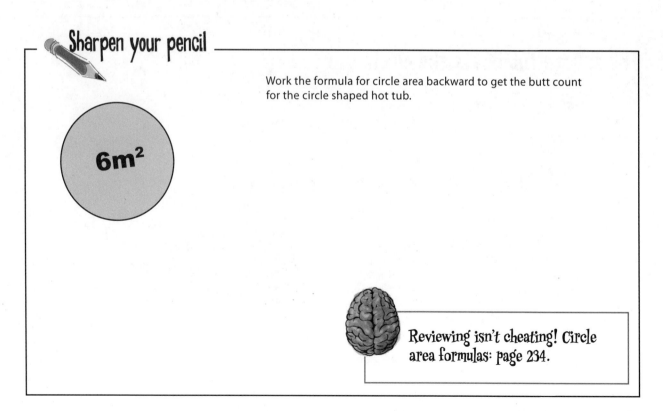

Work the formula for circle area backward to get the butt count for the circle shaped hot tub.

6m²

Reviewing isn't cheating! Circle area formulas: page 234.

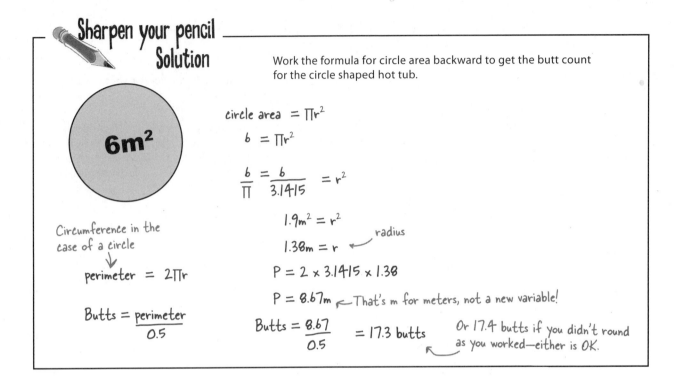

Sharpen your pencil Solution

6m²

Work the formula for circle area backward to get the butt count for the circle shaped hot tub.

$$\text{circle area} = \Pi r^2$$

$$b = \Pi r^2$$

$$\frac{b}{\Pi} = \frac{b}{3.1415} = r^2$$

$$1.9m^2 = r^2$$

$$1.38m = r \quad \leftarrow \text{radius}$$

$$P = 2 \times 3.1415 \times 1.38$$

$$P = 8.67m \quad \leftarrow \text{That's m for meters, not a new variable!}$$

Circumference in the case of a circle
↓

$$\text{perimeter} = 2\Pi r$$

$$\text{Butts} = \frac{\text{perimeter}}{0.5}$$

$$\text{Butts} = \frac{8.67}{0.5} = 17.3 \text{ butts} \quad \text{Or } 17.4 \text{ butts if you didn't round as you worked—either is OK.}$$

The square tub beats the circle tub

So, when it comes to providing butt-space for a 3 cubic meter volume of water, the square shaped hot tub is better than the circle shaped one.

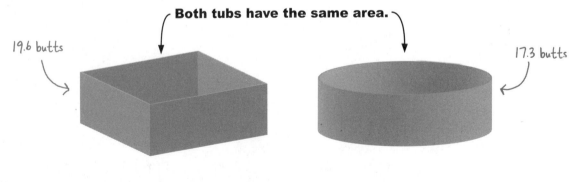

19.6 butts

Both tubs have the same area.

17.3 butts

Both tubs are the same depth.

Two tubs down, five to go

If there were only two tub designs your work would be pretty much done, but there are another five shapes of hot tub to compare in order to find the perfect hot tub for that backstage area at the rock festival.

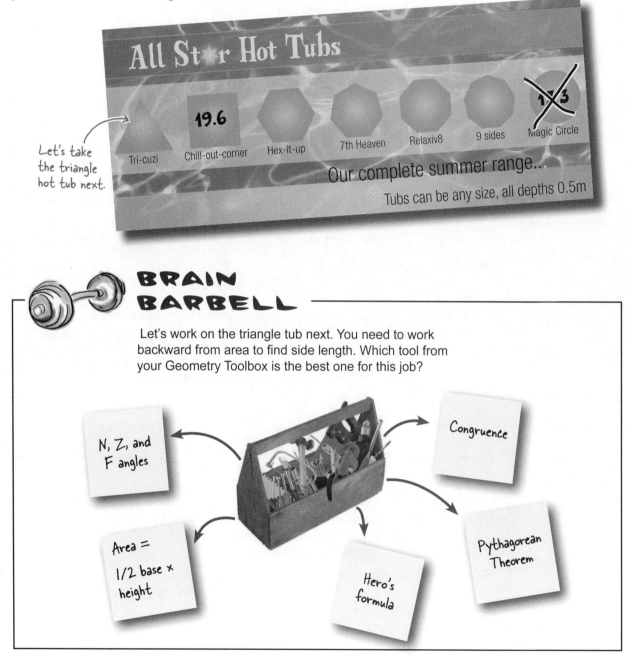

All St★r Hot Tubs

19.6

Tri-cuzi   Chill-out-corner   Hex-it-up   7th Heaven   Relaxiv8   9 sides   Magic Circle

Our complete summer range...

Tubs can be any size, all depths 0.5m

Let's take the triangle hot tub next.

BRAIN BARBELL

Let's work on the triangle tub next. You need to work backward from area to find side length. Which tool from your Geometry Toolbox is the best one for this job?

N, Z, and F angles

Congruence

Area = 1/2 base × height

Hero's formula

Pythagorean Theorem

BRAIN BARBELL SOLUTION

Let's work on the triangle tub next. You need to work backward from area to find side length. Which tool from your Geometry Toolbox is the best one for this job?

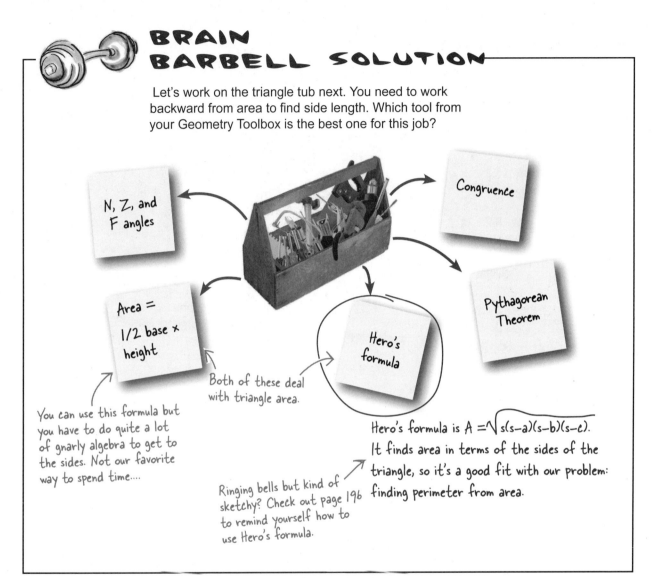

N, Z, and F angles

Congruence

Area = 1/2 base × height

Hero's formula

Pythagorean Theorem

You can use this formula but you have to do quite a lot of gnarly algebra to get to the sides. Not our favorite way to spend time....

Both of these deal with triangle area.

Ringing bells but kind of sketchy? Check out page 196 to remind yourself how to use Hero's formula.

Hero's formula is $A = \sqrt{s(s-a)(s-b)(s-c)}$. It finds area in terms of the sides of the triangle, so it's a good fit with our problem: finding perimeter from area.

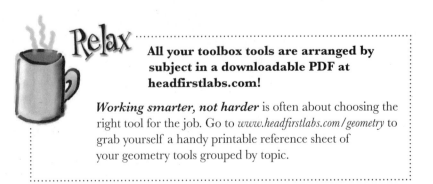

Relax

All your toolbox tools are arranged by subject in a downloadable PDF at headfirstlabs.com!

Working smarter, not harder is often about choosing the right tool for the job. Go to *www.headfirstlabs.com/geometry* to grab yourself a handy printable reference sheet of your geometry tools grouped by topic.

Hot Tub
~~Pool~~ Puzzle

Use the algebra parts in the pool to find a special version of Hero's formula for your equilateral triangle-shaped hot tub. (You should use each part exactly once.)

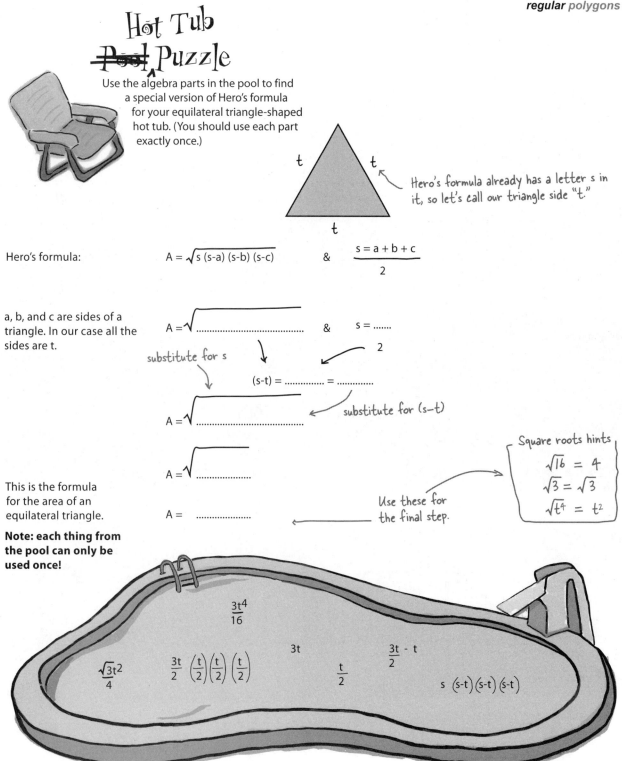

Hero's formula already has a letter s in it, so let's call our triangle side "t."

Hero's formula:

$$A = \sqrt{s\,(s\text{-}a)\,(s\text{-}b)\,(s\text{-}c)} \qquad \& \qquad s = \frac{a+b+c}{2}$$

a, b, and c are sides of a triangle. In our case all the sides are t.

$$A = \sqrt{\rule{4cm}{0pt}} \qquad \& \qquad s = \frac{\dots}{2}$$

substitute for s

$$(s\text{-}t) = \dots = \dots$$

substitute for (s–t)

$$A = \sqrt{\rule{4cm}{0pt}}$$

$$A = \sqrt{\rule{2cm}{0pt}}$$

This is the formula for the area of an equilateral triangle.

$$A = \dots$$

Use these for the final step.

Square roots hints

$$\sqrt{16} = 4$$
$$\sqrt{3} = \sqrt{3}$$
$$\sqrt{t^4} = t^2$$

Note: each thing from the pool can only be used once!

$$\frac{3t^4}{16}$$

$$3t$$

$$3t - t \over 2$$

$$\frac{\sqrt{3}t^2}{4}$$

$$\frac{3t}{2} \quad \left(\frac{t}{2}\right)\left(\frac{t}{2}\right)\left(\frac{t}{2}\right)$$

$$\frac{t}{2}$$

$$s\,(s\text{-}t)(s\text{-}t)(s\text{-}t)$$

Hot Tub ~~Pool~~ ∧Puzzle Solution

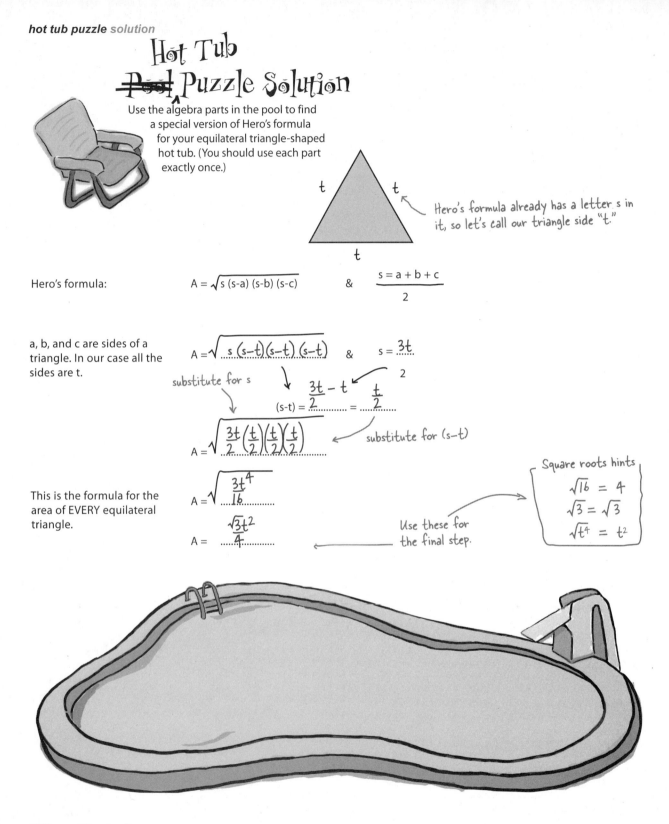

Use the algebra parts in the pool to find a special version of Hero's formula for your equilateral triangle-shaped hot tub. (You should use each part exactly once.)

t t

t

Hero's formula already has a letter s in it, so let's call our triangle side "t."

Hero's formula:

$$A = \sqrt{s\,(s-a)\,(s-b)\,(s-c)}$$ & $$s = \frac{a+b+c}{2}$$

a, b, and c are sides of a triangle. In our case all the sides are t.

$$A = \sqrt{s\,(s-t)(s-t)\,(s-t)}$$ & $$s = \frac{3t}{2}$$

substitute for s

$$(s-t) = \frac{3t}{2} - t = \frac{t}{2}$$

$$A = \sqrt{\frac{3t}{2}\left(\frac{t}{2}\right)\left(\frac{t}{2}\right)\left(\frac{t}{2}\right)}$$

substitute for (s–t)

This is the formula for the area of EVERY equilateral triangle.

$$A = \sqrt{\frac{3t^4}{16}}$$

$$A = \frac{\sqrt{3}t^2}{4}$$

Use these for the final step.

Square roots hints

$$\sqrt{16} = 4$$
$$\sqrt{3} = \sqrt{3}$$
$$\sqrt{t^4} = t^2$$

You've found the formula for the area of an equilateral triangle

Recognizing special shapes is a great shortcut. Being able to spot that a quadrilateral is a square, or that a triangle is an equilateral triangle can save you a lot of work by letting you use extra tools and formulas which apply to that special shape specifically.

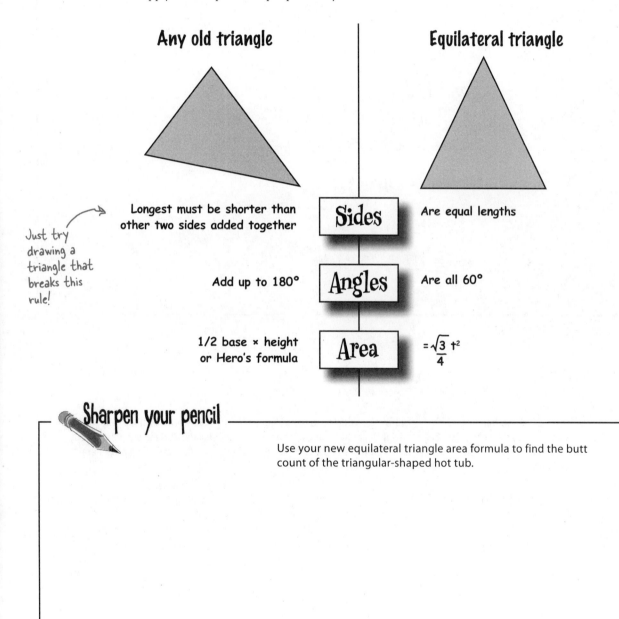

Any old triangle

Equilateral triangle

	Sides	
Longest must be shorter than other two sides added together		Are equal lengths

Just try drawing a triangle that breaks this rule!

	Angles	
Add up to 180°		Are all 60°

	Area	
1/2 base × height or Hero's formula		$= \frac{\sqrt{3}}{4}\, t^2$

Sharpen your pencil

Use your new equilateral triangle area formula to find the butt count of the triangular-shaped hot tub.

Sharpen your pencil
Solution

Use your new equilateral triangle area formula to find the butt-count of the triangular shaped hot tub.

$$\text{Area} = \frac{\sqrt{3}t^2}{4} \longrightarrow 6 = \frac{\sqrt{3}t^2}{4}$$

$$\frac{24}{\sqrt{3}} = t^2$$

Length of each side $\longrightarrow$ $3.72m = t$

$$\text{Perimeter} = 3t$$
$$= 3 \times 3.72$$
$$= 11.17m$$

$$\text{Butts} = \frac{11.17}{0.5} = 22.3 \text{ butts}$$

---

there are no Dumb Questions

Q: Why bother learning another formula? Won't I be covered if I just know Hero's formula and the "1/2 base × height" version?

A: Any time you use Hero's formula to work on a equilateral triangle you're gonna wind up doing all the math we just did anyway. If you can learn the equilateral triangle area formula then you won't need to do all that algebra again!

Q: But what if I wanted to use "1/2 base × height." I thought in geometry it didn't matter how you work stuff out, the answer should be the same?

A: That's true, and starting from 1/2 base × height, you'd have ended up with the exact same formula. You just have to do more work to get there. First you'd need to use the Pythagorean Theorem to find the sides in terms of the height, and then you'd need to do some simultaneous equations to put it all together. It would definitely work but there's a lot more to go wrong.

Q: OK, so this formula only works for equilateral triangles? What about triangles that are nearly equilateral?

A: You could use it to approximate area for a triangle with sides of similar length, but it wouldn't be accurate. It's a good way to check your answer in an exam though. Pick the middle side length, work the equilateral triangle area formula and it should come out in the same ball park. If your two answers are 9 and 10, you're all good; if they're 9 and 50, you know you need to check your work.

Dude, what did you say the circle tub butt-count was again?

Jill: Uh, I think it was like 19 point something.

Frank: No, that's the square isn't it?

Jill: Oh, yeah. Sorry. It's on this other piece of paper.

Frank: And did you keep track of what the dimensions were for the carpenters?

Joe: Yeah, well, uh, not exactly, but I did scribble it down in my notes somewhere so I must have it.…

Frank

Jill

Joe

Hmm. This is getting messy fast. Do you think there's a more efficient way to keep track of the calculations for each hot tub?

Keep track of complex comparisons with a table

There's a whole bunch of stuff you're working out to pick the best hot tub. You need to keep track of side length and angles for the carpenters, as well as butt-space to choose the best tub for the band…and you need to know what calculation goes with which tub! A table can help you make sure all that information is right at your fingertips.

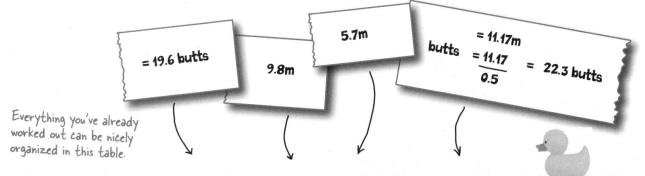

= 19.6 butts

9.8m

5.7m

butts   = 11.17m

$= \dfrac{11.17}{0.5}$ = 22.3 butts

Everything you've already worked out can be nicely organized in this table.

Model	Shape	Polygon	No sides	Perimeter	Butts	Angles	Side length
Tri-tub	△						
Chill-out-corner	▢						
Hex-it-up	⬡						
7th Heaven	⬡						
Relax-v8	⯃						
9 Sides	⬣						
Magic Circle	●						

Exercise

Let's look for some quick answers here!

Based on the work you've already done, you should already be able to fill in all the boxes in this table except for the shaded ones.

Model	Shape	Polygon Name	No sides	Perimeter	Butts	Angles	Side length
Tri-tub	△						
Chill-out-corner	◻						
Hex-it-up	⬡						
7th Heaven	⬣						
Relaxiv8	⯃						
9 Sides	⯃						
Magic Circle	⬤						

These three might make your brain hurt a bit—don't panic, we'll talk more about them later.

Exercise Solution

Let's look for some quick answers here!

Based on the work you've already done, you should already be able to fill in all the boxes in this table except for the shaded ones.

Model	Shape	Polygon	No sides	Perimeter	Butts	Angles	Side length
Tri-tub	△	Equilateral Triangle	3	11.17m	22.3	60°	3.72m
Chill-out-corner	▢	Square	4	9.8m	19.6	90°	2.45m
Hex-it-up	⬡	Hexagon	6				
7th Heaven	⬡	Septagon	7				
Relaxiv8	⬡	Octagon	8				
9 Sides	⬡	Nonagon	9				
Magic Circle	⬤	Circle	∞	8.67m	17.3	Nearly 180°	Nearly 0

If you need help remembering these, check out the "Nonagon" song by They Might Be Giants on YouTube. Goofy but genius.

That's the infinity symbol!

More on these later in the chapter.

Don't worry if these made your head hurt....

That's a pretty table and all, but it looks like you just ran straight into a **DEAD END**. Triangles, squares, and circles are one thing but helloooo—area of a septagon? We don't have formulas for the areas of any of those other shapes. And if we have to work them out one by one this could take forever. By tomorrow? Not gonna happen.

It's a fair point.

On the surface it looks like most of our table is filled in, but we're still a long way from choosing that ideal hot tub.

Time to change our strategy!

Brain Storm

You need to pick a hot tub fast! Write down ANYTHING you can think of that could help you find the area of the 6-, 7-, 8-, and 9-sided tubs. No idea is too crazy. If you run out of space grab some paper. Give yourself at least 5 minutes, and then come and join our brainstorm results over on the next page….

Brain Storm

So what might help us find the area of the 6-, 7-, 8-, and 9-sided tubs? Here are the ideas we came up with, but you might have had different ideas, or extra ideas we didn't think of. The more, the merrier!

We could chop the octagon into rectangles and triangles

We could try that thing we did to work out the pizza area, where we chopped it up into triangle-like slices.

Phone a friend!

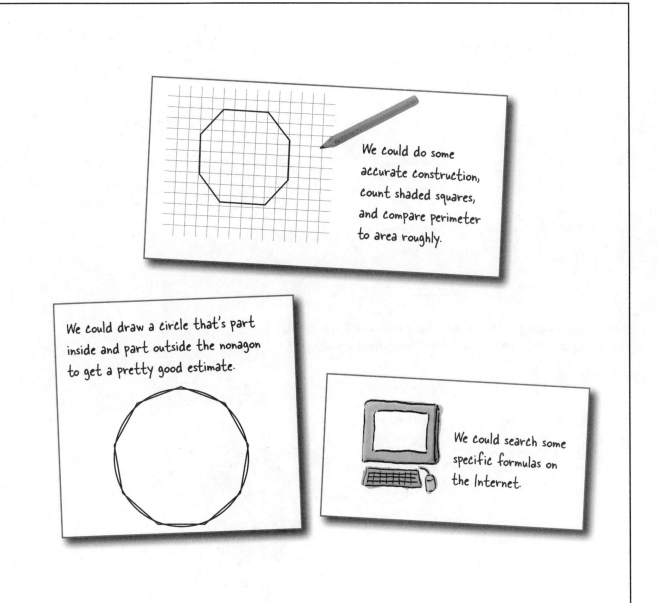

We could do some accurate construction, count shaded squares, and compare perimeter to area roughly.

We could draw a circle that's part inside and part outside the nonagon to get a pretty good estimate.

We could search some specific formulas on the Internet.

What do YOU want to try?

Chop the polygons into triangles

When you investigated the pizza area formula back in Chapter 5, you created a series of polygons by chopping the pizza into triangular slices.

Any regular polygon can be divided into the same number of congruent triangles as the polygon has sides.

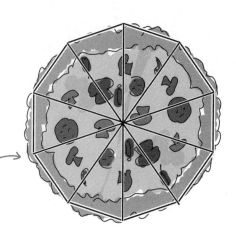

The 10-slice pizza you worked with is actually a 10-sided polygon known as a decagon.

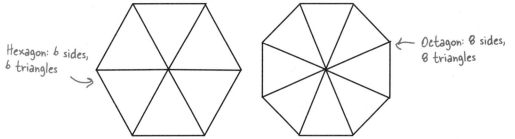

Hexagon: 6 sides, 6 triangles

Octagon: 8 sides, 8 triangles

So, if we know the area of one of those triangles, we can easily find the area of the whole polygon.

A nice idea. Except we don't know anything about those triangles. We don't know base and height or side lengths! Without those we can't find the area. Even Hero's formula can't rescue this one....

That's true. We don't have enough information.

To use either technique we know for finding triangle area we're going to need to know more about these triangles....

What do we need to know about the polygon triangles?

We know that there are the same number of triangles as sides of the polygon, and we know that all the triangles are congruent so we just need to find the area of one. But what else do we need to know to find the triangle's area, so that we can combine them to find the total polygon area?

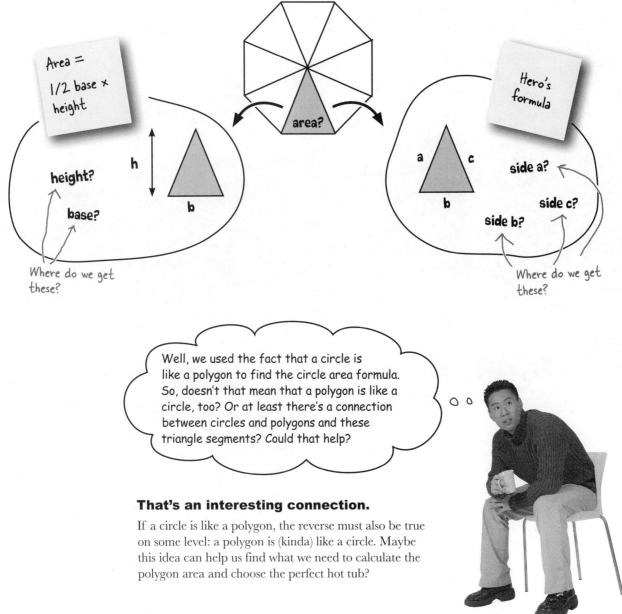

Well, we used the fact that a circle is like a polygon to find the circle area formula. So, doesn't that mean that a polygon is like a circle, too? Or at least there's a connection between circles and polygons and these triangle segments? Could that help?

That's an interesting connection.

If a circle is like a polygon, the reverse must also be true on some level: a polygon is (kinda) like a circle. Maybe this idea can help us find what we need to calculate the polygon area and choose the perfect hot tub?

Geometry Detective

Your mission is to investigate how you could find the missing information needed to use either of the triangle area formulas by using circles fitted to a polygon.

We used these circles to find the triangle centers in Chapter 4. →

The polygons below are drawn with either their circumscribing circle (around the outside, touching every vertice) or their inscribing circle (within the polygon, touching every side).

Draw or sketch on the polygons—we've given you some spares so don't be shy!

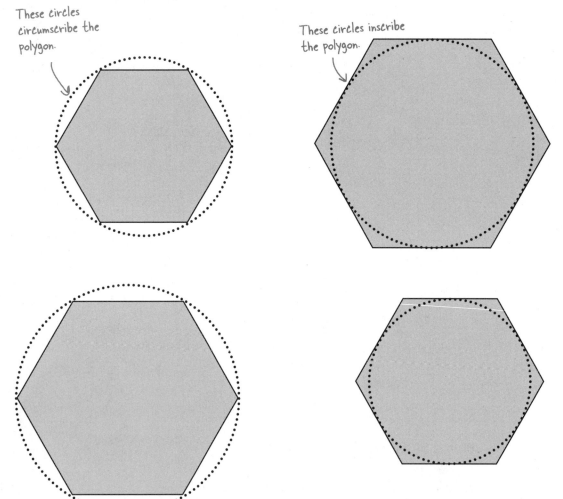

These circles circumscribe the polygon.

These circles inscribe the polygon.

Hint: Start by dividing the polygons into equal triangles. You're looking for base, height, and side length of those.

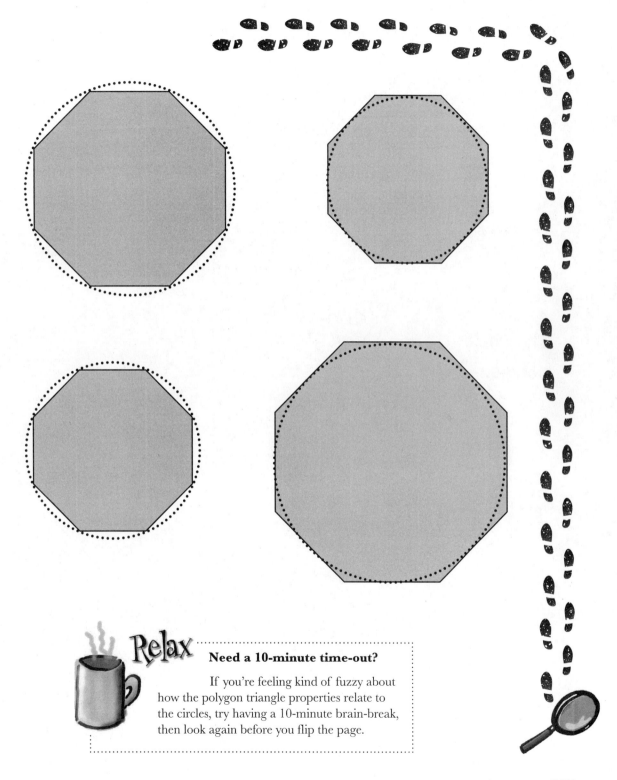

Need a 10-minute time-out?

If you're feeling kind of fuzzy about how the polygon triangle properties relate to the circles, try having a 10-minute brain-break, then look again before you flip the page.

The circles give us the properties we need

The base of each triangle in the polygon is the same as one side of the polygon, and the height and other two sides of the triangle are radii of the circles inscribing and circumscribing the polygon. The radius of the circumcircle (goes around the outside) of a polygon is the polygon's ***circumradius***, and the radius of the incircle (goes around the inside) is called the polygon's ***apothem***.

Radii is just the plural of radius.

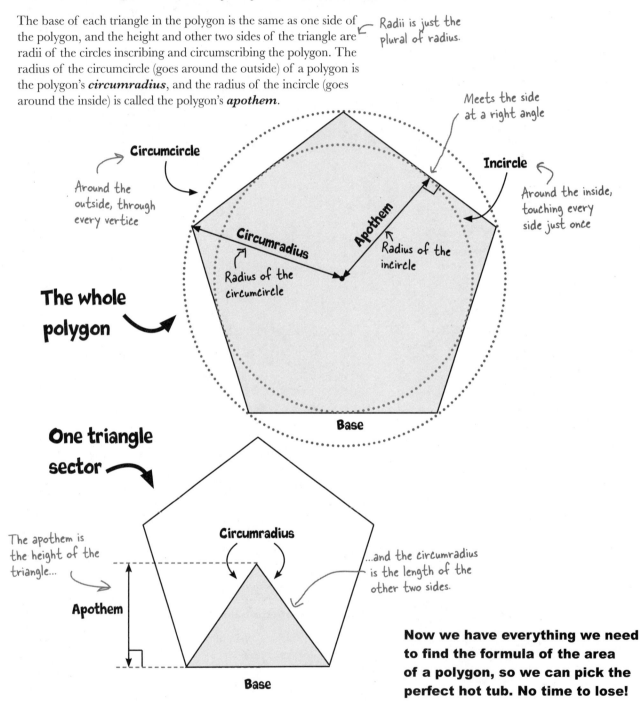

Circumcircle

Around the outside, through every vertice

Meets the side at a right angle

Incircle

Around the inside, touching every side just once

Circumradius

Radius of the circumcircle

Apothem

Radius of the incircle

The whole polygon

Base

One triangle sector

The apothem is the height of the triangle...

Circumradius

...and the circumradius is the length of the other two sides.

Apothem

Base

Now we have everything we need to find the formula of the area of a polygon, so we can pick the perfect hot tub. No time to lose!

Polygon Formula Magnets

Rearrange the magnets to find the general formulas for the perimeter and area of our polygon hot tubs. Use each magnet exactly once.

(Hint: it looks a lot like one of the intermediate stages of the circle area formula!)

Think carefully about which area formula you're using.

Area of one triangle 'sector' = [ ] [ ] × [ ]

Total area = [ ] [ ] × [ ] × [ ]

Combine these.

Perimeter = [ ] × [ ]

Total area = [ ] [ ] × [ ]

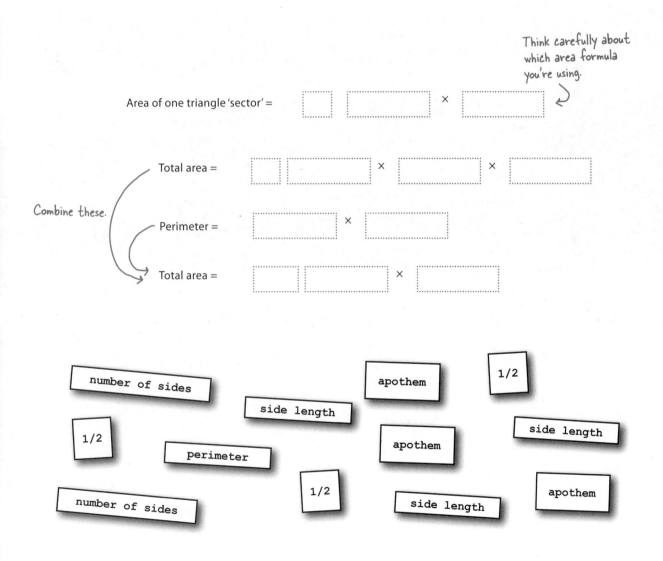

number of sides

side length

apothem

1/2

1/2

perimeter

apothem

side length

1/2

side length

apothem

number of sides

Polygon Formula Magnets Solution

Rearrange the magnets to find the general formulas for the perimeter and area of our polygon hot tubs. Use each magnet exactly once.

(Hint: it looks a lot like one of the intermediate stages of the circle area formula!)

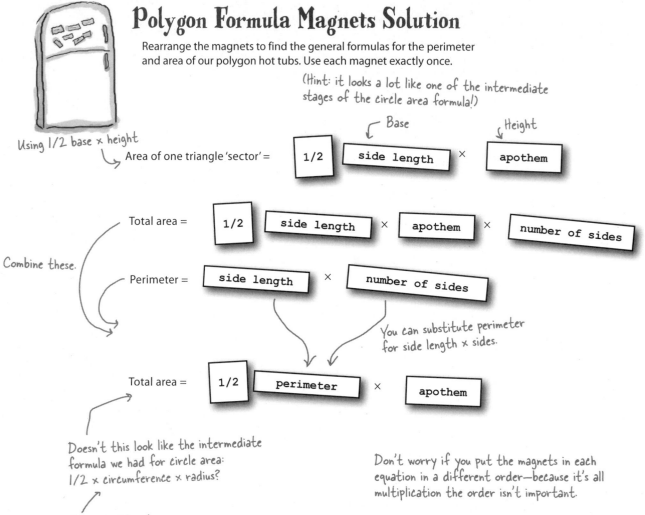

Using 1/2 base × height

Area of one triangle 'sector' = | 1/2 | side length | × | apothem |

⌐ Base
↓ Height

Total area = | 1/2 | side length | × | apothem | × | number of sides |

Combine these.

Perimeter = | side length | × | number of sides |

You can substitute perimeter for side length × sides.

Total area = | 1/2 | perimeter | × | apothem |

Doesn't this look like the intermediate formula we had for circle area: 1/2 × circumference × radius?

See page 230 for a reminder if you're fuzzy.

Don't worry if you put the magnets in each equation in a different order—because it's all multiplication the order isn't important.

BULLET POINTS

- The **apothem** joins the center to the middle of a side, bisecting it.

- The apothem meets the side at a **right angle**.

- The **circumradius** joins the center to a vertex.

- The circumradius **bisects** the internal angle.

Polygon area = 1/2 perimeter x apothem

When you add up all the areas of the triangle sectors in a polygon you end up with something pretty neat: 1/2 perimeter × apothem.

Often you'll be given the apothem, or you might have the circumradius and the side length and then you can use the Pythagorean Theorem to find the apothem.

> OK, great. So that gives us a formula that we can use for all the hot tubs, whether they've got 6, 7, 8, or 9 sides. That **is** cool.

Joe: Yeah, cool, whatever. Except we still need to find a bunch of apothems to use it!

Frank: Relax, I'm not sure we actually do.

Jill: How come?

Frank: Well, the whole way we found that circle area formula was about adding more and more sides to the polygon until it resembled a circle—yeah?

Joe: Yeah. And?

Frank: So how many butts fit in the circle tub?

Jill: Not many! It's the least we've found. Triangle was the most, square was pretty good.

Frank: Exactly! That's not just a detail—that's a trend!

What do you think the trend is?

Which hot tub are you leaning toward?

Frank

Jill

Joe

More sides = fewer butts

Look again at that intermediate circle area formula compared to the polygon area formula you just created:

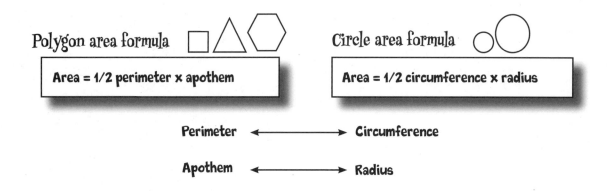

Polygon area formula

Area = 1/2 perimeter x apothem

Circle area formula

Area = 1/2 circumference x radius

Perimeter ⟷ Circumference

Apothem ⟷ Radius

A circle really is just another polygon with an infinite number of sides. At the other end of the scale you've got the humble triangle, the lowest polygon of all. In between those two sit every other regular polygon you can imagine.

So far, we've found that the most butt-space comes from the hot tub with the fewest sides, and the least butt-space comes from the hot tub with the most sides. It's a trend!

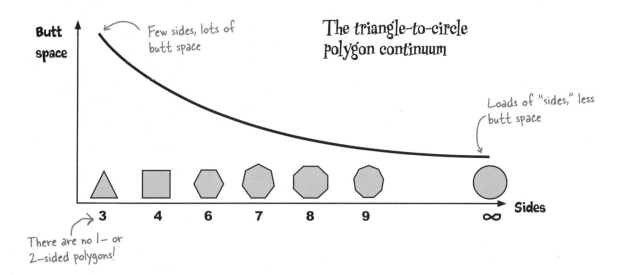

Butt space

Few sides, lots of butt space

The triangle-to-circle polygon continuum

Loads of "sides," less butt space

3   4   6   7   8   9   ∞   Sides

There are no 1- or 2-sided polygons!

Rock stars—high maintenance?

Uh oh. The band is getting demanding. They've seen the All Star Hot Tubs brochure lying around and think there's a couple of designs that look pretty uncool.

At least you don't have to pick out all the red M&Ms from their snack bar...

BAND called RE: hot tub.
NO SQUARES or TRIANGLES
Too boring....

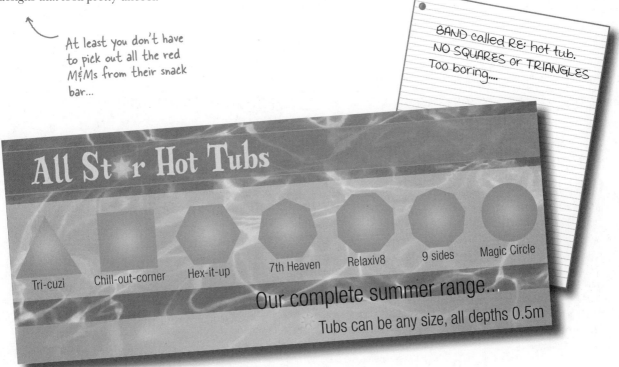

All St★r Hot Tubs

Tri-cuzi   Chill-out-corner   Hex-it-up   7th Heaven   Relaxiv8   9 sides   Magic Circle

Our complete summer range...

Tubs can be any size, all depths 0.5m

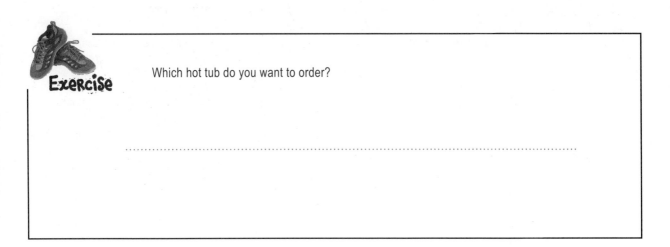

Exercise

Which hot tub do you want to order?

....

Exercise Solution

Which hot tub do you want to order?

The HEX-IT-UP tub with six sides.

there are no
Dumb Questions

Q: A circle is just a polygon with infinite sides? Have you gone a bit philosophical or what?

A: Some math geeks get a bit upset about this definition of a circle, but if you could draw enough sides then you'd be approaching a circle. The whole point of infinity is that you never get there, but you get close enough for this to be the basis of computer graphics.

Q: Computer graphics? Polygons? How does that work?

A: To show 3D curvy worlds, all video games rely on triangles, which build into polygons, which fool our eyes into believing we're seeing circles and curves.

Q: Ah! Yeah, when I've zoomed in real close to something in a game sometimes I've seen that it's made out of flat shapes.

A: That's it—exactly.

Q: Circumradius sounds a lot like radius, but in the comparison of the area formulas you compared apothem to radius. That's kind of confusing!

A: Actually a regular polygon's circumradius is sometimes just referred to as its radius. But if you think about it, a circle's apothem and its circumradius can both be equal to its radius (if the apothem is going right to the edge). So when we compared the apothem and the circle's radius, we were comparing an apothem to an apothem!

Q: Does the polygon area formula work for all regular polygons? Even squares and triangles?

A: The polygon area formula works for any regular polygon—whether it's got 3 sides or an infinite number! But you've got some much sharper tools in your tool box for finding square and triangle areas, and they'll always be quicker and easier to use, so it's best to keep this one for regular polygons with five sides or more.

Q: Regular polygons only? This area formula won't work with irregular polygon?

A: Sadly not. Irregular shapes are harder to deal with anyway. When you split them into triangles the triangles aren't congruent. Very high maintenance. Luckily they don't come up in exams much!

Q: Apothem. Let me guess, that means in an algebra version of the formula that would be an "a"? But we're already using "a" for area—why couldn't they have named it something with a different first letter?

A: Good question. We don't have an answer—except that if you remember your formulas as words rather than just letters then it doesn't matter so much. If you're sitting a test and you want to demonstrate that you know the formula, write it out in words and then make it clear which letters you're picking as stand-ins. Whether you use a, A, s, or t for something is less important than what it means. It's also easier to work out whether a formula applies to a problem when you remember your formulas as words and not just letters.

Great tub choice!

Your pick of the six-sided hot tub has stirred up
excitement among all the bands booked for the
festival—they think it's gonna be amazing. And you
didn't even have to do any more math to choose the
hexagon tub, since you already found that handy trend.

> Yeah—great, except that the
> carpenter wants the dimensions and
> angles of this fantasy six-sided tub
> that you've picked. Please don't tell me
> that chopping it up into triangles is
> going to find those as well....

Actually, it might.

It's not a bad suggestion at all. We have plenty of triangle-
related tools in our toolbox, so when we're faced with a
problem we can't already solve, looking for ways to make it
about triangles is a great starting point. Let's give it a go.

BRAIN BARBELL

We know we can divide the hexagon tub into six congruent
triangles. From this, can you find:

1) The total of all the angles of those "sector" triangles.

2) The angle that the carpenters need to build the wood
frame for the hot tub: the internal angle of the polygon.

A handy
hexagon to
doodle on
↓

The carpenters
want this
internal angle.
↙

BRAIN BARBELLSOLUTION

We know we can divide the hexagon tub into six congruent triangles. From this, can you find:

1) The total of all the angles of those "sector" triangles.

2) The angle that the carpenters need: the internal angle of the polygon.

Add up ALL the angles in the triangles...

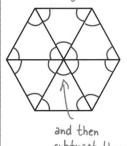

and then subtract these six in the middle.

1) Angles in a triangle add up to 180°

There are six triangles so the total angles are: 6 x 180° = 1080°

2) The six angles in the middle add up to 360°

So the total of all the internal angles is the total angles minus the six angles in the middle = 1080° – 360° = 720°

It's a regular polygon, so the angles are equal = $\frac{720°}{6}$ = 120°

Internal angles of a polygon follow a pattern

A polygon can always be divided into the same number of triangular sectors as it has sides. And every triangle has angles totaling 180°. That's the key to finding the internal angles of a polygon:

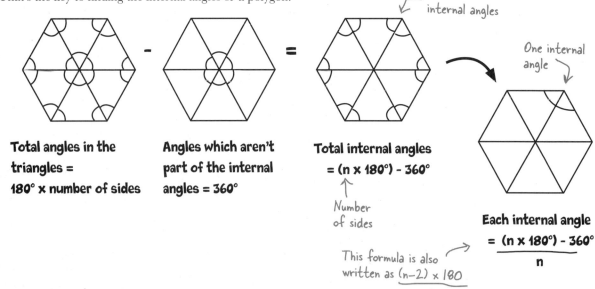

All the internal angles

One internal angle

Total angles in the triangles = 180° x number of sides

Angles which aren't part of the internal angles = 360°

Total internal angles = (n x 180°) - 360°

Number of sides

This formula is also written as $\frac{(n-2) \times 180}{n}$

Each internal angle = $\frac{(n \times 180°) - 360°}{n}$

But what about dimensions?

Phew! The only thing left on your to-do list is to tell the carpenters
what length each side of the hot tub is.

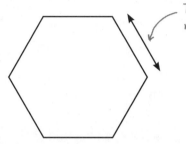

This is what the carpenter
needs to know.

BRAIN
BARBELL

The total area of the hot tub is 6 square meters. Use what
you know about the hexagon hot tub angles to find the length
of each side of the hot tub. (Hint: it helps if you split it into
those sector triangles!)

Hint—look back at page 293 if you're not sure where to start....

BRAIN
BARBELL SOLUTION

The total area of the hot tub is 6 square meters. Use what you know about the hexagon hot tub angles to find the length of each side of the hot tub.

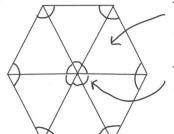

These 6 triangles are the same (congruent). Each one must have an area of 1 square meter.

The angles in the middle add up to 360°, so this angle here must be 60°.

An isosceles triangle has two equal sides, and two equal angles.

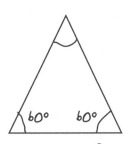

Each internal angle is 120°—and we know the triangle is ISOSCELES because the side lengths are both equal to the circumradius of the polygon, so the other angles must both be 120°/2 = 60°.

Each 1 sq meter triangle is EQUILATERAL.

Any triangle with three 60° angles is equilateral.

So we can use the equilateral triangle area formula!

$$\text{Area} = \frac{\sqrt{3}t^2}{4}$$

t is side length.

Area of each triangle is the total polygon area divided by the number of triangles.

$$1m^2 = \frac{\sqrt{3}t^2}{4}$$

$$\frac{4}{\sqrt{3}} = t^2$$

$$2.31 = t^2$$

Use a calculator!

Length of each side.

$$1.52m = t$$

It's time to relax in the hot tub!

You've earned it. Not only have you picked the perfect hot tub to please both the environmental engineers and the bands, you've even managed to sort out the exact angles and dimensions for the carpenters to get ready to install it immediately!

Along the way you've created a bunch of new tools for your Geometry Toolbox, and discovered how lots of your tools can work together to solve really complex problems.

Now—has anybody seen those rubber ducks?

Room for 18 butts, including yours

Exactly 3 cubic meters of warm, bubbly water

Nice! We love this festival. We should make it an annual thing....

CHAPTER 7

Your Geometry Toolbox

You've got Chapter 7 under your belt and now you've added regular polygons to your toolbox. For a complete list of tool tips in the book, head over to www.headfirstlabs.com/geometry.

Perimeter =
Number of sides
× Side length

Area =
1/2 × Perimeter
× Apothem

Circumradius:
radius of the circumcircle

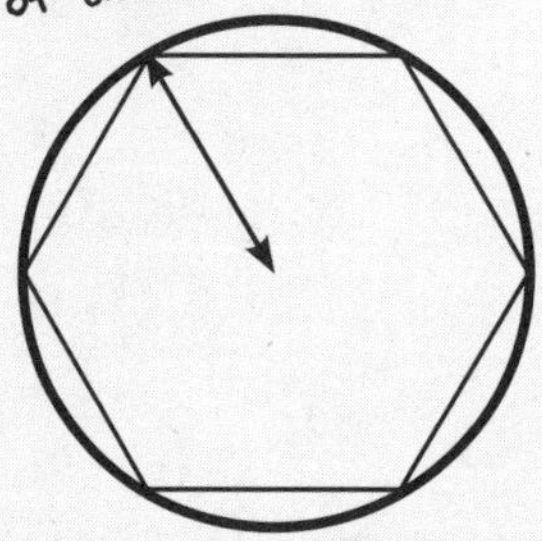

Equilateral
triangle area =

$$\frac{\sqrt{3}\ t^2}{4}$$

Apothem:
radius of the incircle

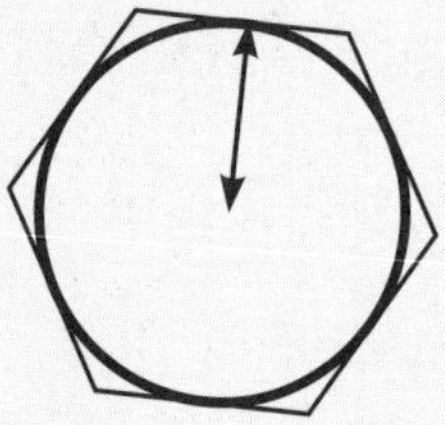

Internal angle =

$$\frac{(n \times 180°) - 360°}{n}$$

(n is number of sides)

BULLET POINTS

- The **apothem** joins the center to the middle of a side, bisecting it.

- The apothem meets the side at a **right angle**.

- The **circumradius** joins the center to a vertex.

- The circumradius **bisects** the internal angle.

Leaving town...

It's been great having you here in Geometryville!

We're sad to see you leave, but there's nothing like taking what you've learned and putting it to use. You're just beginning your geometry journey and we've put you in the driver's seat. We're dying to hear how things go, so *drop us a line* at the Head First Labs website, **www.headfirstlabs.com/geometry**, and let us know how geometry is paying off for **YOU**! Next stop? *Head First 3D Geometry*!

Index

Symbols

∞ (infinity symbol), 298

π (Pi symbol), 214–216

Numbers

3D shapes, volume of, 278–280

3-sided shapes (see triangles)

4-sided shapes (see quadrilaterals; squares)

6-sided shapes (see hexagons)

7-sided shapes (see septagons)

8-sided shapes (see octagons)

9-sided shapes (see nonagons)

A

accurate construction, 108–112

acute angles, 10

acute triangles, 131, 148

adjacent angles, 9

altitude line, 144–145, 148

angles, 3
    acute angles, 10
    adding, 7–10
    adjacent angles, 9
    complementary angles, 9–10, 12–13, 47
    cutting into smaller angles, 7, 46
    equal angles, 30
    of equilateral triangles, 176

    of parallelograms, 246, 251
    of quadrilaterals, adding, 34–37, 46
    in regular polygons, 313–314, 318
    right angles, 6, 11, 14, 46, 106
    spread angle of speakers, 170–172, 174
    of straight lines, 11, 46
    supplementary angles, 17–18, 47
    of triangles, adding, 20–22, 46
    of triangles, estimating, 175–180
    vertical angles, 19, 22, 47

apothem of regular polygon, 306, 308, 312, 318

arc of circle, 219–222, 234

area
    of circles, 223–232, 234, 288, 310
    compared to perimeter, 250, 309–310
    of congruent triangles, 159
    of equilateral triangles, 290–294, 318
    of kites, 253–259
    of parallelograms, 238–246
    of rectangles, 156–162
    of regular polygons, 280–282, 284, 302–310, 312, 318
    of scalene triangles, 196
    of sectors of a circle, 232, 234
    of squares, 286
    of trapezoids, 260–265
    of triangles, 162–167, 202, 290–294

B

ballistics example (see homicide example)

bases of trapezoid, 262

bisecting isoceles triangles, 144–146

bounding rectangle, 157–163

C

center of triangle, 181–188

centroid of triangle, 182, 186–188, 202

circles

    arc of, 219–222, 234

    area of, 223–232, 234, 288, 310

    circumcircle of regular polygon, 304, 306

    circumcircle of triangle, 183

    circumference of, 214–216, 234, 288

    comparing to regular polygons, 303–306

    diameter of, 208, 234

    incircle of regular polygon, 304, 306

    incircle of triangle, 182

    radius of, 208, 234

    as regular polygons, 298, 312

    sectors of, 210, 226–228, 232, 234

    similarity of, 77, 100

circumcenter of triangle, 183, 186, 202

circumcircle of regular polygon, 304, 306

circumcircle of triangle, 183

circumference of circle, 214–216, 234, 288

circumradius of regular polygon, 306, 308, 312, 318

complementary angles, 9–10, 12–13, 47

congruent shapes, 81–82, 89, 100

congruent triangles

    area of, 159

    dividing regular polygons into, 302

crime scene example (see homicide example)

cubes, volume of, 278

D

diameter of circle, 208, 234

drawings

    accuracy of, 5, 8, 30

    accurate construction of, 108–112

    equal angles marked on, 30

    parallel lines marked on, 39

    right angles marked on, 6, 46

E

8-sided shapes (see octagons)

equal angles, markings for, 30

equilateral polygons (see regular polygons)

equilateral triangles

    angles of, 176

    area of, 290–294, 318

    center of, 188

    perimeter of, 294

    (see also regular polygons)

examples (see homicide example; hot tub example; lawn service example; myPod example; pizza example; rock festival example; skate ramp example)

F

F patterns, in parallel lines, 40, 47

factor trees, 121, 122

4-sided shapes (see quadrilaterals; squares)

fractions, compared to ratios, 94

H

Hero's formula, 196–200, 202, 290–292

hexagons, 298, 312–316

    (see also regular polygons)

homicide example, 2–44

    angle calculations from Benny to bullet in wall, 4–23

    angle calculations from Charlie to car to bullet in wall, 28–44

hot tub example, 274–317

    angles in hot tub, 313–314

    area of hot tub, 282

    perimeter of hot tub, 284–310

    side lengths of hot tub, 315–316

    volume of hot tub, 278–280

hypotenuse of triangle, 131, 138

I

incenter of triangle, 182, 186, 202

incircle of regular polygon, 304, 306

incircle of triangle, 182

infinity symbol (∞), 298

integer solutions for right triangles, 136

iPod etchings example (see myPod example)

irrational numbers, 216

isosceles trapezoids, 267, 269

isosceles triangles, 144–145, 175–180, 269

K

kites, 253–259, 268, 272

L

lawn service example, 236–271
    area of lawn, 237–246, 253–259, 260–266
    perimeter of lawn, 248–251, 259, 266

legs of triangle, 131, 138, 141

line segments, 6

lines, 6
    accuracy of, in drawings, 5, 8
    as angles, 11, 46
    cutting into smaller angles, 14–16
    parallel lines, 39–43, 47
    perpendicular lines, 6, 106
    (see also legs of triangle; perimeter)

M

markings on drawings (see drawings)

median, 182, 202

methods of solving problems, 108–112, 116–118

myPod example, 50–98
    shape calculations for logo design, 78–88
    shape calculations for mountain design, 53–66
    size calculations for logo design, 90–98
    size calculations for mountain design, 68–77

N

N patterns, in parallel lines, 40, 47

9-sided shapes (see nonagons)

nonagons, 298
    (see also regular polygons)

non-integer solutions for right triangles, 136

O

obtuse triangles, 131, 148

octagons, 298
    (see also regular polygons)

orthocenter of triangle, 182, 186, 202

P

parallel lines, 39–43
    F, Z, and N patterns for, 40, 47
    markings for, 39
    vertical angles and, 40, 47

parallelograms, 269, 272
    angles in, 246, 251
    area of, 238–246
    compared to trapezoids, 270
    perimeter of, 248–251

perimeter, 152–155, 202, 251
    of circles (see circumference)
    compared to area, 250, 309–310
    of parallelograms, 248–251
    of regular polygons, 277, 284, 310–312, 318
    of squares, 286–287
    of triangles, 294

perpendicular lines, 6, 106
    (see also right angles)

Pi symbol (π), 214–216

pizza example, 206–233
    comparing pizza sizes and prices, 223–232
    comparing slice sizes and prices, 207–211
    pepperoni crust prices, 212–221

polygons, 275
    (see also regular polygons)

problems, methods of solving, 108–112, 116–118

proportional shapes, 73–76

Pythagoras/Pythagorus (see Pythagorean Theorem)

Pythagorean Theorem, 131–135, 138, 141–146,
    148, 194

Pythagorean triples (Super Triangles), 146, 148

Q

quadrilaterals, 268, 272
    corner angles of, adding, 34–37, 46
    types of, 268–270
    (see also squares)

R

radius of circle, 208, 234

range of speakers, 170–172

ratios, 73, 92–98

rectangles, 156–162, 269, 272

regular polygons, 275–276
    apothem of, 306, 308, 312, 318
    area of, 280–282, 284, 302–310, 312, 318
    circles as, 298, 312

circumradius of, 306, 308, 312, 318
comparing to circles, 303–306
congruent triangles in, 302
internal angles in, 313–314, 318
perimeter of, 277, 284, 310–312, 318
side lengths of, 315–316
(see also equilateral triangles; squares)

rhombuses, 268, 270, 272

right triangles, 110–115, 119–131
    integer vs. non-integer solutions for, 136
    Pythagorean Theorem and, 131–135, 138, 141–146, 148

right angles, 6, 106
    markings for, 6, 46
    pairs of, 11, 14
    (see also perpendicular lines)

rock festival example, 150–201
    area of field shapes, 156–168
    fence prices for field shapes, 151–154
    location for drinks stall, 181–189
    screen for obstructed view area, 190–200
    speaker range and spread, 170–180

rope swing example, 139–147

S

scalene triangles, 194–198

scaling shapes, 70–76, 121–122
    (see also similar shapes)

sectors of circle, 210, 226–228, 232, 234

septagons, 298
    (see also regular polygons)

7-sided shapes (see septagons)

shapes
    area of (see area)
    congruent (see congruent shapes)

perimeter of (see perimeter)
scaling (see scaling shapes)
similar (see similar shapes)
(see also specific shapes)
shooting example (see homicide example)
similar shapes, 58–63, 67, 77, 89, 100
proportional scaling of, 73–76
ratios for, 92–98
right triangles and, 120–122
(see also scaling shapes)
6-sided shapes (see hexagons)
skate ramp example, 104–147
multiple ramps, 132–135
perpendicular uprights, 106–131
rope swing, 139–147
sketches (see drawings)
speakers example, 170–180
spread angle of speakers, 170–172, 174
squares, 268
area of, 286
perimeter of, 286–287
similarity of, 77, 100
(see also quadrilaterals; regular polygons)
straight lines (see lines)
Super Triangles (Pythagorean triples), 146, 148
supplementary angles, 17–18, 47

T

3-sided shapes (see triangles)
3D shapes, volume of, 278–280
trapeziums, 270

trapezoids, 262, 267, 269, 272
area of, 260–265
bases of, 262
compared to parallelograms, 270
isosceles trapezoids, 267, 269
triangles
altitude line for, 144–145, 148
angles of, adding, 20–22, 46
angles of, estimating, 175–180
area of, 162–167, 202, 290–294
bisecting, 144–146
bounding rectangle for, 157–163
centers of, 181–188
equilateral triangles, 176, 188, 290–294, 318
hypotenuse of, 131, 138
isosceles triangles, 144–145, 175–180, 269
legs of, 131, 138, 141
medians of, 182, 202
perimeter of, 294
right triangles (see right triangles)
scalene triangles, 194–198
similar triangles, 58–63, 67, 73–76

V

vertical angles, 19, 40, 47
volume of 3D shapes, 278–280

Z

Z patterns, in parallel lines, 40, 47